T0226940

Troubleshooting
Switching Power Converters

Troubleshooting
Switching Power Converters

A Hands-on Guide

By

Sanjaya Maniktala

AMSTERDAM • BOSTON • HEIDELBERG • LONDON
NEW YORK • OXFORD • PARIS • SAN DIEGO
SAN FRANCISCO • SINGAPORE • SYDNEY • TOKYO

Newnes is an imprint of Elsevier

Newnes is an imprint of Elsevier
30 Corporate Drive, Suite 400, Burlington, MA 01803, USA
Linacre House, Jordan Hill, Oxford OX2 8DP, UK

 Recognizing the importance of preserving what has been written, Elsevier prints its books on acid-free paper whenever possible.

Library of Congress Cataloging-in-Publication Data
Maniktala, Sanjaya.
 Troubleshooting switching power converters : a hands-on guide / By Sanjaya Maniktala.
  p. cm.
 Includes bibliographical references and index.
 ISBN 978-0-7506-8421-7 (hardcover : alk. paper) 1. Electric current converters.
 2. Semiconductor switches. 3. Switching power supplies. I. Title.
 TK7872.C8M26 2007
 621.31'7—dc22   2007023364

British Library Cataloguing-in-Publication Data
A catalogue record for this book is available from the British Library.

ISBN: 978-0-7506-8421-7

For information on all Newnes publications
visit our Web site at www.books.elsevier.com

Printed and bound by CPI Group (UK) Ltd, Croydon, CR0 4YY

Transferred to Digital Print 2011

Contents

Preface

A few weeks ago, I found myself groping for some sorely needed inspiration. I was even questioning the very *need* for a book on this particular subject. So despite spending five years in the general vicinity of the legend behind it, I finally went and bought myself the "other" book—*Troubleshooting Analog Circuits* by Robert Pease. I am glad I did, because it ended up fulfilling both my requirements—I not only learned that his book is truly is an inspiring resource (certainly something to chuckle your way through coach class with pretzels and coffee in hand), but also that it *isn't* about Power Conversion. Take, for example, that famous picture on its cover. The hopelessly entangled object you see there (the creation, not the creator) apparently served as a historic V-F (voltage-to-frequency) converter circuit. But really, it would never pass muster as even a basic *switching* converter. Breadboards, for one, are kryptonite to switching converters. If you really think about it, all that that picture so aptly conveys is exactly what you *shouldn't* ever be attempting to do in power. Reassuringly, even that book itself recognizes that "switch-mode regulators [are] a whole new ball-game."

So it was surprising for me to learn that the analog troubleshooting book was originally intended to be just a *single chapter* of a much larger volume on the topic of *switching power converters*. Maybe that project just slipped through the cracks of time. Perhaps it was too difficult a venture to undertake—a hypothesis somewhat supported by the fact that there is still virtually no other book out there on this topic (cheers to the book you're holding, by the way!). But I also tend to believe that if it were published in the format it seems to have been originally conceived in, it could well have turned out to be quite misleading—for reasons very similar to that on the cover of the analog book. For a while, that had me seriously thinking: "whose bright idea was *that*?" Then I realized that back in those days, Power Conversion was still in its infancy. Who knew what lay ahead?

In the early 1970s, the Intel 8080 microprocessor was dazzling engineers around the world with its blazing computational speed of 2MHz! "Digital" became the anthem for a new generation of EE graduates. Virtually every prospective hire our company interviewed in that decade (and the next), when asked what he or she would like to do for the company, said without a blink—"*Microprocessor!*" I think I probably got to do Power only because I was not considered good enough for all the really "good stuff." Though a little later I think things must have soured for some of the laggards involved in this digital race, individuals and companies alike. And in the resulting reverse pileup, somehow everything *non-digital* suddenly got crowned "analog."

I wonder how that happened. Catchy phrases such as "the power of analog" may have tickled the collective imagination, but at best they were just oxymorons (quite appropriately with a "moron" in it!). Because, despite all their *apparent* familiarity and similarity, "Analog" and "Power" are actually strange bedfellows. You just can't club them together in one grand compendium titled "Everything you *didn't* want to know about *digital* circuits . . . and *weren't* afraid to ask." Yes, though both analog and power have certain commonalities, in that they are both essentially "non-digital," the similarity ends right there. The distinguishing characteristic of modern switching power converters is that their sharp *edges* of current and voltage, driven by an almost *mystically endowed inductor*, generate so much high-frequency content that all the hitherto painfully learned rules of the game tumble helplessly to the *ground* (quite literally so).

"Analog or Power, what's in a name?" you may well ask. One practical problem arising from that lack of dichotomy is that in many companies, senior managers suddenly arrive almost unheralded on the scene, possessed with the driving vision and unfettered desire to steer the huge switcher business to new heights—based solely on the credentials that they know everything about *op-amps*. In a recent case inside a major analog company based in Santa Clara, California, the newly appointed senior VP of the two successful product lines "Power Management" and "Portable Power" pulled aside his staff to ask almost incredulously, "Why do you call our switchers 'simple?'" He was convinced that somehow this totally degraded the product—maybe like simple burgers, or simple minds. Notwithstanding the glaringly obvious fact that *that* had in fact been the most recognized brand name of the company for the last ten years. Should we call them "Performance Switchers?" Or "Blazing Switchers?" How about "Complex Switchers?" Or maybe "Fiery Switchers?" (Now that would make engineers run to buy 'em!) I was told a hush fell in the conference room right after. A few careerists were still nodding their heads in unbridled awe, but no one there was going to be the one to tell the Emperor that he really needed a (good) tailor now.

Looking further "down" the food chain (or should I say "up"), till just a few years ago, Power was a niche market that most engineers *didn't* want. Today it is a niche market that most *can't* do. It has turned out to be one thing to write a brilliant paper on the subject, or even put together some heady course material for, quite another to get *on the bench* and really *build* a converter that works. Remember, you also need to make it exceedingly reliable, and it costs peanuts in the bargain. That's the name of the game. At least one reason for this unfortunate situation is that very few engineering schools out there still teach any significant amount of power electronics, especially about switch-mode power supplies (SMPS). It's a double-E without the SMPS thrown in. Unfortunately, that is just not enough. Companies are scrambling to try and hire graduates from the handful of schools that specialize in power. The situation really needs to be rectified quickly to cope with the steeply escalating demand for trained graduates.

The field of Power has gotten not only critically important, but incredibly complex too. No longer can a wannabe expert get by based solely on the rather astute observation found deep within the Analog Troubleshooting book—"if you stand on a big soapbox and rant and holler, people will often think you know what you are talking about. They stop looking for mistakes . . . and that's a mistake." Yes, I do remember that thunderous pounding of the copy machine in Building D punctuated by some of the choicest epithets I have heard this side of the Pacific. That *was* truly unmistakable! Luckily, I have quickly learned from my own experiences that peers who help you find mistakes *before* you make them, are your best friends in the business. Those who help you find them *after* you make them are your well-wishers. And those who never say a word, before or after, are your *real* mistakes. Make no mistake about that!

In the scientific and engineering community of today, we are all becoming increasingly subject to the same level of critical examination and crosschecks as anyone else out there. No longer can we hide behind the mistakes of others, or rant and rail to misdirect attention away from our own mistakes. We must therefore learn to come clean whenever necessary, and also to do that as quickly as possible to avoid any significant fallout from our erroneous actions or advice. For that is what will ultimately drive progress—ours and *theirs*. We must likewise also start demanding the same standards from everybody associated with the field. No longer should we take shiny flyers, media presentations, or slick online tools at face value. The bar *must* be raised, and very soon. Power happens to be so tricky an area that not only are there plenty of honest mistakes abounding (and I make my fair share everyday), but this field also offers plenty of exotic buzzwords, scary equations, and impressive-sounding "trade-off claims" to take refuge under—if that's what you *intended*. Therefore, even vendors, for example, however high and mighty they may be, need to be subject to the same level of *intense scrutiny* that we are expecting to fall on ourselves. That is just engineering the way I understand it. If we don't, I feel the costs will simply proliferate and grow for all concerned. The world is shrinking with every passing day, and in fact we are already deeply connected. Can we really afford to pretend otherwise?

Finally, it is time for me to say goodbye. Three books, you'll agree, are enough! There just can't be another one. This book was, therefore, my last chance to tell you some of the *stories* behind the experiences. I was also hoping to make it interesting and memorable in the process, make you feel like *you* lived through it yourself. Because that way, I figure,

you are less likely to ever forget the *technical* learnings attached to the stories, either.
So as I put my pen down once and for all, I thank you for your tremendous support always.
I truly hope you not only learn from this book, but also enjoy it—as much as I did while
writing it! I expect this one to be considered rather blunt in places, too, but I promised you
the truth and this is it.

—Sanjaya Maniktala

Acknowledgments

I would like to thank my mentor, VP and friend Doctor G. T. "Doc" Murthy for his abiding inspiration many years ago in Bombay. He set an example for me when I was still young, which I find myself trying to emulate even today. Among other things, I learned from Doc what *engineering* is all about. At least, what it *really* ought to be. Without trying to sound pompous here, the fact is that I realized clearly that no person, or organization, can ever rise above the basic threshold of *integrity* they abide by. Otherwise, the value of any data, claim, or product emanating from that source will always be suspect. Genuine mistakes or errors are of course entirely natural and excusable. We are human after all. But the only proof that the mistake was indeed involuntary, should be judged by the fact that its proponent went back, rechecked the data, and then if an error was noticed, backtracked immediately without further dalliance—and publicly so. No true engineer (or engineering organization) should ever want to, or need to, hide in any possible way ("catch me if you can," or "huh, what did you say, duh?"). Further, no ISO quality standard out there can ever hope to replace that basic and necessary engineering attitude. This book rather strongly reflects that personal belief of mine. Blame Doc for that!

Many thanks are also due to some great people I met recently who sorted out a lot of personal issues of mine. In particular, thanks to Travis Meek, a young and extremely thorough attorney at Fairchild's immigration firm, Pierce Atwood LLP, for his awesome support during the writing of this book. Without his help, I would have been terribly distracted for sure. And, of course, I would also like to thank Professor Shafiq Rahman at Allegheny College.

I have some special readers out there who have taken the trouble of expressing a lot of appreciation for my previous books. They probably don't realize how much that means to a writer like me. In that context, I would like to thank Robert Rauck and Paul Mathews in particular. I would also like to thank Dipak Patel at my old company, National Semiconductor.

Thanks are also due to Carol Mohr and Jeff Freeland working for Elsevier, for helping create a very professional look for this book, in a very short time (yes, I was late by two months in delivering the manuscript).

Many thanks to Disha, Aartika, and Munchi, for yet again spending months without me at home, while I struggled feverishly to get this all done. I had promised them last year (during the writing of my second book), that that would be the last I would ever write. This time, I have at least been consistent in telling them that this is again the last one. But I have a feeling they stopped believing me years ago!

Thinking Power

"Houston, Wir haben ein Problem"

The Germans and the Swedes spell it the way we'd spell it. The Spaniards, Portuguese, Russians, Greeks, and Italians add an "a" after it. The Dutch and the French include an extra "e" in it somewhere. But we all do seem to agree on one thing—that it is basically a *problem*. And a pretty universal one, too.

That is also the scope of this book—specifically, *problems* in switching power supplies. It is therefore full of seemingly nonstop quibbling about trivial details. Loaded with relentless griping and not-so-subtle reminders of past failures. Bursting with an almost compulsive agenda for harping only about the defects surrounding us.

But it really is necessary! Because finally, we are in that shadowy realm bordering reality and fantasy, where everything that can possibly go wrong *will*, along with everything that was considered impossible just seconds ago! That is the world of switching power supplies, as I know it.

However, to set expectations correctly: no single book (or diatribe in this case) can ever hope to detail all the countless possibilities of failure. What follows is necessarily only representative. Though it should certainly help spark off a search in the right direction, and very quickly.

At least I am hoping it does, because otherwise, we really do have a problem!

Practice and Theory: Two Sides of the Same Coin

One of the first things we need to be aware of while troubleshooting switching converters is that we must develop an ability to look at a practical problem through the eyes of supporting theory, and vice versa. Ultimately we all have to learn to keep transitioning almost seamlessly between these two modes (glitch-free), questioning our rationale every thousandth of an inch (i.e., mil) of the way. Because, especially in this area, theory and practice have become so meticulously intertwined that neither can survive on its own. For example, in theory you could have a 100MHz power converter switching away even as of today. But clearly, its very existence on earth has been limited thus far to a fraction of that frequency, by what are purely practical concerns. Similarly, you could come up with an excellent layout plus thermal management solution—something to dazzle your professor with. But if you are shaky about the very basic principles governing the switching topologies, you could well find yourself working fruitlessly in your lab into the dead of night for months, if not years.

While on the bench, what you really need planted very firmly on your head at all times is a diehard, practicing *engineer's* hat. Neither an ordinary technician's hat, nor a pure academician's hat, can work entirely on its own. It's become one thing to write a brilliant exposé on the subject, quite another to build something that works (properly). Recognizing that fact, this book will try to guide you more effectively by highlighting the theoretical principles underlying the myriad practical aspects of the field.

What you won't find here is some mindless do-it-yourself technician's guide with easy step-by-step troubleshooting instructions, like: check the voltage on Pin 5, check if the cap marked C21 is leaking fluid, replace the charred resistor R16, and so on. That just won't do justice to the complexity surrounding this remarkable and exciting field. And neither can that approach ever hope to be comprehensive enough. We just have to learn to attack every new problem we face (and there are many more coming at us as we speak), with the force of underlying principles, logic, and experience (hopefully sometimes someone else's experience!). That is what we will be trying to build up in these pages. Because those are the only tools we can really count on to get us through in the long run.

It must also be pointed out that since we don't have the luxury of introducing a lot of the basic concepts or terminology used here, you may want to read up on some good basic material first. Quite naturally, I recommend my most recent book *Switching Power Supplies A to Z*. That one was designed to bring you up to speed as quickly as possible, so you can at last be at par with your blazing 4MHz switcher!

Leave Your Past Perceptions Behind

As we take our first steps into the world of troubleshooting, we must consciously attempt to leave our perceptions, presumptions, and past intuition behind us. We must start afresh, and learn to analyze whatever lies ahead, with the due diligence of an unremarkable, but *assiduous* engineer. Because nothing really is what it seems to be.

I know you are already thinking that that is not necessary, at least as far as you are concerned. So I am first offering you a simple challenge in two steps. It is not even a question about Power. It's not even concerning any fancy AC analysis. It's a simple DC-bias op-amp question that I often ask job candidates during interviews. You will be surprised how many experienced engineers end up scratching their heads (and sometimes even mine when I am not looking).

Challenge 1: Take a look at Figure 1-1. In Step 1: what is the voltage on the output pin of the op-amp? Simple, it is 2V. Correct! Now coming to Step 2, your gut instinct probably screams out *Vout* = 4V. But that is wrong. To find out why, read the explanation within the figure. In fact, a few years ago, I even posed this challenge to the foremost IC designer (now Design Manager) of the Power Management group of an analog company I worked at. He too first got it wrong, then smiled rather sheepishly. We must realize the importance of putting pen to paper and double-checking the seemingly obvious.

Challenge 2: In Figure 1-2, we have two identical capacitors. One is charged to 10V. Then it is connected directly across the other capacitor. What are the final voltages across the capacitors?

The "correct" answer is 5V. But actually, I am going to leave you guessing about the exact reason for this (if you don't already know). There are several websites dedicated to this topic, so please Google freely. It is interesting to learn that half the energy originally

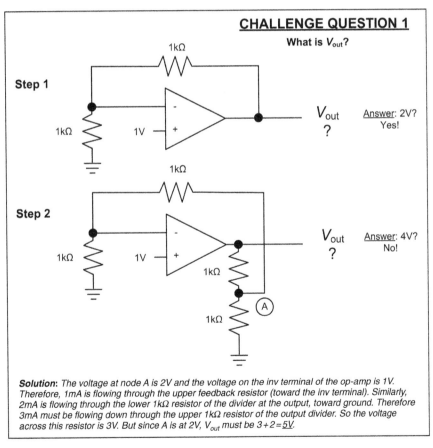

Solution: *The voltage at node A is 2V and the voltage on the inv terminal of the op-amp is 1V. Therefore, 1mA is flowing through the upper feedback resistor (toward the inv terminal). Similarly, 2mA is flowing through the lower 1kΩ resistor of the divider at the output, toward ground. Therefore 3mA must be flowing down through the upper 1kΩ resistor of the output divider. So the voltage across this resistor is 3V. But since A is at 2V, V_{out} must be 3+2=5V.*

Figure 1-1 Test Your Intuition (Challenge Question #1)

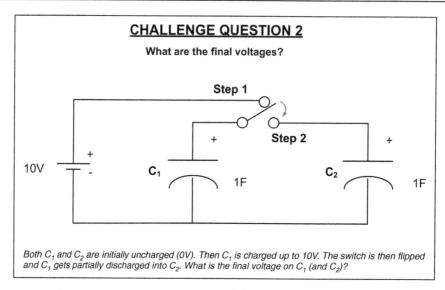

Figure 1-2 Test Your Intuition (Challenge Question #2)

present inside C_1 is just lost somewhere! Where, why, and how? I must add that recently a senior colleague of mine disagreed with the traditional resolution of this paradox. He said, "Show me!" And that is a good thing, because however obvious it seems to be, we must ultimately take it to the bench and try it out. So we did! We charged up a big electrolytic capacitor to 10V, and then immediately connected it across an identical uncharged one. The final voltage? About 5.7V, not 5V. I still don't understand that! Though I suspect it was probably a result of the tolerances and mismatching of the capacitors involved.

Choose Your Friends Carefully

I have become increasingly wary of some of the slick media and marketing presentations around us. As engineers, we are likely to spend months troubleshooting parts, often putting our careers on the line. It is therefore getting increasingly important for us to realize that some companies actually seem to be counting on ignorance. Yes, it is true that most media-persons, for example, don't know much about Power Conversion. That's not really surprising or embarrassing, especially since many of us working in the field often realize how little we really know! But as a result of this, it has become almost easy for vendors to dish out glossy flyers, perhaps with accompanying fancy Flash/HTML tricks, hoping people will buy into their claims without any further questions. And they usually do!

I saw one advertisement recently claiming that a new IC was the ultimate improvement because its board had gone from 65 components to 40 components. This marketing presentation was from a popular vendor of high-voltage monolithic ICs for AC-DC applications. My version of the ad can be seen in Figure 1-3. I came up with some questions. 1) The new board was missing a huge heatsink. How can the heatsink disappear? If it's for the same application, the dissipation in the switch will be roughly the same. 2) The old board had some large blocky EMI filter components, but there was no EMI filter on the new board. Well, since the integrated switcher IC is not soft-switching or slew-controlled, the EMI would be roughly the same in the new board, and it would still need this filtering. 3) The old board had a standard fuse, which was reduced in the new board to a tiny radial fuse. Well, the size of the fuse does not depend on the switcher IC being used. All of these issues were clear from a glance at their boards, but the advertising jargon made it hard to see through the murk to the real issues. I still think it

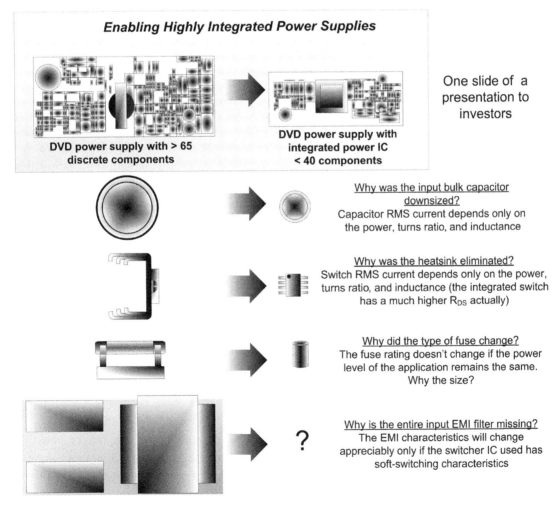

Figure 1-3 What Some Marketing Presentations Will Not Tell You

is somewhat unfortunate the company even felt the need to do this, especially because *their parts are really good enough to deserve much better.*

We need to start questioning and analyzing in great detail what a particular vendor is really promising (versus not promising) and what the vendor is publishing (but perhaps not publicizing or highlighting). All of this will impact our design sooner or later, perhaps far more than we imagined early on. We will be punished for the slightest ignorance or presumption on our part, about their part. So rather than spend valuable time looking for Apps fixes (application-level solutions) and flimsy Band-Aids, it will save us a lot of grief if we learn to do our homework and pick the ICs (and remaining components) correctly in the very early stages of design, thereby knowing exactly what we are really getting into. We will find that certain vendors will hold up admirably to our in-depth engineering scrutiny and analyses, but some others, such as in the case of my example of the misleading advertisement, will simply fall apart. The truth is, the final measure of a company is not related to either the size of its tiny evaluation boards, or its overblown media presentations. For example, there are massive companies like TI (Texas Instruments) that are fairly well-respected for their straightforward, diligent datasheets and related product information. Such companies have, over the years, become clearly identifiable by peers as nothing but serious engineering entities. Unfortunately, there are also several small to medium-sized "TI wannabes," some of whom even go as far as to suggest they have by now become the very *sight and sound of information*, and thereby represent nature itself. My experience is we really need to be on guard the moment we hear exceptionally crafted slogans emanating from a certain vendor. Dazzling catchphrases like "the power of analog" are there solely to dazzle. Perhaps for a good reason too! One that you don't want to discover the hard way. The signs are already there that the company is probably spending far more energy in marketing than on its expected core competence, i.e., *engineering.*

Yes, it is always heartening to learn that there are in fact several vendors that do care about not misleading their customers, and are therefore relatively up-front about the capabilities of their devices. So for example, if their device is not a good fit for a particular application or has a specific weakness in that application, they would rather just say so (hoping you will come back to them later when the time is ripe). Fair enough!

Another side of the coin we call marketing hype is the unfortunate problem of deceptive or junk ICs (I sometimes refer to them as *silicone* chips). There is a surprising proliferation of such products out there, and rarely do their manufacturers ever recall them or even admit to their "oops." As a prospective customer, if you had known beforehand that the IC you picked for the job was just a sandy quagmire lying patiently in wait for you, maybe you could have avoided the headache of trying to debug it. I am not merely talking about parts with slender manufacturing tolerances, or quaint little bugs, because that is actually fairly common. I am talking about parts that are *conceptually* wrong to start with. Power is unforgiving—if you defy its basic principles, you can be certain it will come back to hound you one fine day. And when that happens, no book such as this one can even hope to be on your side.

Let's take a look at some of these classic "bloopers" of the semiconductor industry (the British insist on calling them "bloomers," but then, they also drive on the wrong side of the road!):

- **A PFC controller IC with *peak* current mode control:** These were the earliest ones from Microlinear Corp., over a decade ago. Though the devices were industry firsts at that time, the company had not realized that the only way to really correct the shape of the *average* input current waveform drawn from the wall socket was to use *average* current mode control, not *peak* current mode control. Unitrode Corp. (now part of TI) got that right, and so despite a slow start, quickly became the market leader. Microlinear faded into extinction. But certainly not without causing avoidable grief to a lot of engineers who must have thought the company knew what it was doing. Unfortunately, it didn't! This historical blunder was in fact brought to my notice by a visiting Microlinear sales representative himself, when I was working in Germany many years ago. Come to think of it, perhaps I wouldn't be the one hurriedly recommending hiring him though!

- **A *Flyback* controller IC with a *maximum duty cycle of 100%*:** Such a product was actually released in 2000 by a major analog IC manufacturer and is probably still haunting engineers around the world. A Flyback topology depends on a *minimum guaranteed off-time* to get the energy in the transformer to freewheel into the output, and to also be able to reset the leakage inductance. So if the control circuit, in an effort to get the output to rise, keeps the switch ON permanently (100% duty cycle), there is in fact zero time available to actually deliver that energy into the output—a classic Catch-22 situation. Therefore it is no surprise that this very part has generated some of the highest number of queries from hot and bothered engineers around the world on the company's own online discussion forums. Without going into detail, I can tell you that I made the problem known to the company the very same year it was released.

- **A modern-day *Push-Pull* topology with *voltage mode* control:** This was released by a major analog vendor in 2003. Many of us luckily know by now that you really need *current* mode control, not *voltage* mode control, to avoid core saturation and imbalance in the Push-Pull topology. No surprise therefore that the only evaluation board (also called an eval board or EVB) the company itself could come up with was one featuring the Half-Bridge topology, not the Push-Pull. And we do agree that the Half-Bridge topology is in fact well-suited for voltage mode control. But surprisingly, nobody is apparently being warned *not* to try and build a Push-Pull converter with this supposed Push-Pull IC.

■ **A *1MHz* "Emulated Current Mode Control"** *Buck* **IC:** Released with much fanfare by a well-known analog manufacturer in 2006, this IC has a maximum duty cycle of only 50% at 1MHz—all because it needs a whopping 500ns off-time to sample the diode current (to internally create the emulated "DC portion/pedestal" of the switch current ramp). (Incidentally, most other manufacturers are trying to get from 98% to 100% maximum duty cycle in their Buck ICs!). The result is that this IC's frequency has to be reduced to 300kHz for even the most basic applications. Now I can see why the part offers an *adjustable* frequency. But didn't you initially think that was a nice "feature" to have, one that was worth paying some extra money for? Further, if Silicon Valley's notorious grapevine is to be believed, it seems that what was originally about to be released just a few months prior, was an "emulated current ramp" but *without its accompanying DC portion*. Looks like the entire team had been through innumerable design reviews, and were totally convinced they had just built a killer current-mode IC (i.e., one without the noise plaguing traditional current-mode control). Till a last-minute Bode plot showed up the classic LC double-pole response so characteristic of voltage-mode control! Of course the first thing they did after that was to turn around and fire their senior applications engineer. Bad guy! Then they scrambled to somehow save the show. And that's how the critical 500ns off-time seems to have crept in. Incidentally, a few months later, at another company, I found myself staring at an interesting resumé on my table—it was from that fired engineer. He too had enough off-time now!

You might argue that all this probably just goes to show that the solutions to most problems in switching power are really very tricky and even the experts fumble sometimes. That is in fact true, but the system works as long as these experts also display the necessary expertise to quickly recognize their mistakes, pull such products from the marketplace, and redesign them. *Waffles can't replace wafers.*

To illustrate this point, in the last case mentioned above, the company is often seen nowadays rather innocently redirecting attention away from the devastating 500ns minimum off-time of this family of parts to the fairly good 80ns of minimum on-time. But these are two entirely different things really, their impact depending specifically on the application on hand! At the very least, prospective customers should have been warned well in advance about the potential impact of the 500ns minimum off-time on their applications.

Unfortunately, the situation gets even worse. In all the accompanying datasheets the very first hint of this major product limitation occurs deep inside the electrical characteristics tables, in which the 500ns number appears as a mere "Typical" (which most power supply designers know really amounts to nothing). Thereafter, in most of this family's datasheets, the only other indication of this problem occurs several pages later, in the form of a brief paragraph calling out for a low operating frequency. In fact, for the 50 to 500kHz part (released in late 2005), the datasheet skips even mentioning this necessary lowering of frequency. Though it is interesting that the design example inside the datasheet somehow correctly picks the frequency as 300kHz. The reason? It says: "Operation at 300kHz was selected for this example as a reasonable compromise for both small size and high efficiency." Note that that is just a catchall statement applicable to every single converter ever built. So why is 300kHz considered an optimum in this particular case? Why not 500kHz? Actually, it really has nothing to do with either the efficiency or size here. 300kHz is a must only because of the 500ns off-time limitation. A spade is clearly not a spade anymore! It has evolved into an "advanced agricultural apparatus crafted out of materials derived from Stellar nucleosynthesis" (the origin of iron and nickel as per Wikipedia). In fact, this particular datasheet makes absolutely no mention of the 500ns limitation over its entire 22 pages. Correction—there is an exception, in the form of a rather disingenuous statement tucked away deep on Page 13: *When operating with a high PWM duty cycle, the Buck switch will be forced off each cycle for 500ns to ensure that the bootstrap capacitor is recharged.* That's even more surprising, because I have never seen any part from any vendor, that couldn't top up a measly 22nF bootstrap capacitor in 50 to 100ns, if not less. Why does it take 500ns to do the job here? I suddenly realized that this rather creative explanation now also provides the company the opportunity to call the 500ns minimum off-time by a unique name: "*forced* off-time." My guess here is that this interesting shift of nomenclature will make you instinctively think that somehow this 500ns was introduced on purpose and therefore presumably for a good cause. I suspect that calling it a minimum off-time would be clearly perceived as a major limitation by most customers. Marketing has clearly evolved into an art form! All I hope for is that the company's IC designers become equally inventive soon and fix the off-time issue after firing their lying Apps guy/manager.

I remember one of my previous Marketing Directors (in 2005), who expressed this rather aptly (and also candidly, though in a moment of obvious indiscretion):

> "Things have changed a little, but there are times where we will decide to release a part where we know there is a wart . . . we won't say the wart is a wart . . . we may just say there is a bump here. . . . The key is, when you are standing here, [we say] look that way, when you are standing there, [we say] look here. . . . That way they don't see the wart. . . . Ha Ha Ha!"

(He sure had a memorable guffaw, that guy. I will remember him for that at least!)

Finding Solutions that Converge

The task is daunting—a typical switching power supply, for all its apparent simplicity (at least in terms of the typical number of components on the board), probably constitutes one of the most difficult challenges of modern electronics. A simple symptom such as an overheated or unreliable transistor switch may rack the brains of even an experienced engineer for weeks, if not months. The problem could lurk almost anywhere—in the control circuitry, in the magnetics design/selection, in the choice and/or quality of the remaining components, or even in the overall assembly (the PCB, soldering, grounding, heatsinking, etc.). Further, all these potential causes and their respective symptoms may themselves be keenly interrelated. Getting one out of the way may simply exacerbate or unmask another problem, much like a balloon that you press in at one end, only to find it bulging out at the other. Ultimately, you have to learn the art of pressing this balloon in from all sides simultaneously. Only then can you assume the solution has truly converged.

Picture yourself as a determined panther slowly circling in on your unsuspecting prey. It will take time, patience, and plenty of experience to get it right every time. The ultimate challenge in the art of troubleshooting switching power supplies is to be able to solve a problem *without causing another*. And that is much easier said than done! Due to the complex interplay of several engineering disciplines involved in a real-world switching power converter, there is almost nothing we can change at any given point inside it without creating some sort of ripple effect into other aspects of its performance and/or reliability.

The Ripple Effect in Power Supplies

A young engineer (Yongyi) goes home one evening, pleased that he has finally licked the problem. His UC3842-based "baby" is now able to deliver its fully intended power. The very next morning he is having the following exchange with Mr. Ng, his rather overheated Boss.

Boss Ng: Hey Yongyi, I just walked past your bench! I turned down the variac just a tiny little bit. Now I see some really strange pulsing. I think the output voltage ripple has gone up too.

Yongyi: Oh really? I just changed a *tiny* ceramic capacitor—the one connected between the timing capacitor pin of the IC and the current sense pin. In fact I only reduced it from 47pF to 33pF! Or maybe 22pF, I can't remember exactly, but a *verrry* minor change, sir.

Boss Ng: Why did you do that?

Yongyi: Oh, because it was not reaching *full* load sir. I didn't have a sense resistor of a slightly lower value. So I tried changing that capacitor. And it seemed to work like a dream.

Boss Ng: You must have been dreaming all right. Because I think you have gone and ruined the slope compensation! You should never, never, never fool around with that capacitor in this current mode control IC. If you were just a little smarter, you would have tried paralleling a larger resistor across the sense resistor to adjust its value, but you had to go and do just that!

(Well, if dear Yongyi had been any smarter, would he have been working there in the first place?) The technical details of this career-threatening episode are presented in Figure 1-4. But the general message is more important to heed here. We will always run into situations

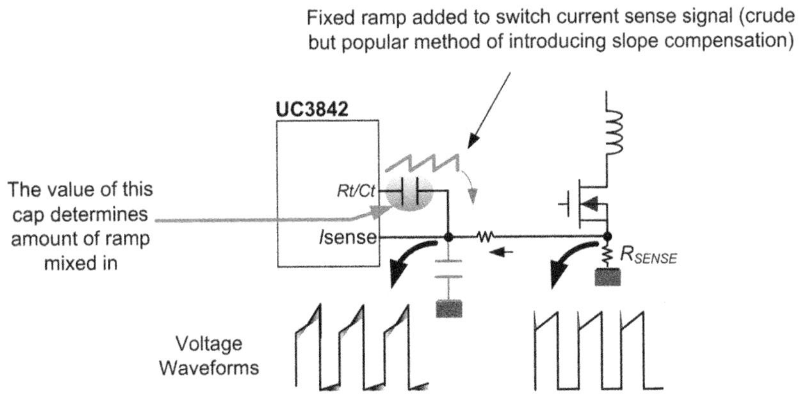

Fixed ramp added to switch current sense signal (crude but popular method of introducing slope compensation)

UC3842

The value of this cap determines amount of ramp mixed in

Rt/Ct

Isense

R_{SENSE}

Voltage Waveforms

Figure 1-4 Changing the Capacitor Instead of Tweaking the Sense Resistor Can Create Instability

like the above, where a certain problem responds to both Fix 1 and Fix 2. If Fix 1 solves the original problem, but in the process creates a new (and unacceptable) problem, we need to go with Fix 2 and not Fix 1. Otherwise the solution cannot be considered to have converged.

Experience Does Count; No Ifs, Ands, or Buts about That

Whenever we have identified a potential fix, we need to test a whole lot of other stuff just to confirm that the overall performance is still acceptable and in spec (within specification). Of course, the number of distinct parameters we may need to check in the process is roughly inversely proportional to our years of hands-on experience. An experienced engineer will likely know almost beforehand where the secondary impact of the change may be felt. And so he or she would quickly hunt it down and test it, not only to ensure that the converter has remained in spec, but that the desired design margins have been maintained. On the other hand, a rookie engineer should always be extremely wary of declaring a quick victory. He or she really needs to spend a much longer time evaluating and validating the fix before going home to his or her noodles, naan, or bratwurst. Ultimately, he or she might need to go through several such fixes and/or iterations before being sure the solution is valid, and has really converged.

An alternative that is readily available to the young engineer (and to the company) at such critical moments is to have a senior (more experienced) engineer intervene and help quickly evaluate the proposed fix. Surprisingly, very few companies seem to have realized the importance of deliberately injecting experience into a project at the right time and at the right place. Resource management to them simply means provisioning for enough multimeters and oscilloscopes in the lab.

But that was in fact the procedure we followed quite religiously in the Singapore lab I worked in several years ago. I guess to outsiders we were certifiably paranoid. But perhaps we alone knew that "only the paranoid survive" (after former Intel CEO, Andrew Grove's book, by the same name). We realized we were faced with the benumbing responsibility of designing and building millions of mission-critical power supplies each year, for a very demanding computer manufacturer (well-known even today for its alternative operating system architecture and portable music players). We had learned from bitter experience that the tiniest oversight can come back to haunt you (frankly, not an atypical occurrence in any large-volume production environment). It was never enough to just get one or two prototypes working on the bench, document the fix, freeze the design, and breezily move on.

Note that an obvious alternative to the problem of "injecting experience" is simply to hire only 20+ years experienced engineers for each and every project. In principle, no further help is ever required (except perhaps from the company's creditors)! I suspect some companies may have tried this approach too, but for some strange reason don't seem to be around to tell their side of the story.

Another approach, wildly popular in some analog companies, is to have an almost completely unsupervised gaggle of junior engineers darting around in the dark for weeks if not months, falling over each other's mistakes. Surprisingly, that is the very situation created by the typical "one-person one-project" assignment structure so common to many major semiconductor companies. A cynical interpretation of that strategy could be divide and conquer. But mostly I think it's just an utter lack of imagination. The Boss, who we graciously assume is still "technical" and therefore could have helped at such a critical juncture, seems to be too busy traveling around business class to some remote destination along the Pacific Rim. But lest we judge him too harshly, remember he is out there looking at the *bigger picture*.

Many years ago I visited a major US manufacturer's design and production facility in remote Youghal, off the coast of the Republic of Ireland. I was absolutely thrilled to meet up with a certain "Peter," their seniormost engineer, who was by then probably the best-known power supply designer in all of Europe. And for good reason! I already knew that even giant companies like Siemens AG (where I worked at that time in Leipzig) were busy analyzing and tearing apart Peter's historic workhorse rectifier designs (rectifiers are switching power supplies meant explicitly for telecom applications). So, over some timely Guinness (in Ireland that's approximately 25 hours a day, in Germany 26!), Peter confided in me that he really had no assigned project anymore. Rather he was their chief troubleshooter (or firefighter, as I think he put it). He told me that was the very reason he was so satisfied working there, because "very few companies realize the importance of a role like that." Touché.

I can vouch for that too! I used to get constant nudges/hints from my Boss while working at this analog semiconductor company not too long ago, "Sanjaya, for the salary I pay you, I can hire at least two guys fresh out of college, and I guarantee they would be kicking butt in less than two months." Come to think of it, they did have many products to prove their point. Except for one small lingering detail—with surprising frequency it turned out to be the *customer's* butt (no ifs, ands, or buts about that).

Never Ignore a Problem Until It Is Too Late

We may have just become guilty of trying to build and test the era's most highly efficient converters in the most inefficient manner. Usually no one ever needs to know about any of this. But it can certainly affect you, one fine day. Particularly if you have not been looking out for the signs. As a smart, wary, prospective buyer, you should try hard to look past any glitz being thrown at you. And then if you are lucky, you might suddenly see unveiled, in full glory (however briefly), the sight and sound of disinformation. For example, you might notice some almost unimaginable mistakes have been made, such as an entire family of integrated so-called simple Buck switcher ICs that can barely work up to half their rated load if the required operating duty cycle happens to exceed 50%. To me it would be almost mind-boggling to imagine that not one engineer in that company ever attempted for years on end to just set up one evaluation board, with say an 18V input and a 12V output, then to short the output and release it. Because they would have seen it for themselves, the perpetual motorboating and loss of output recovery that many customers have slowly discovered. Sadly, such a family of parts was actually released by a major analog manufacturer in 1998. Then suddenly in 2003 their datasheets were hurriedly rev'ed up with a long-winded 410-word disclaimer rather thoughtfully buried deep inside their datasheets (where most companies expect engineers not to delve). Surprisingly, even as of today (2007), the company's online tools actually tell you to just go right ahead and use the devices up to their full rated load without the slightest hesitation, irrespective of your duty cycle. Makes you think. Don't they ever read their own datasheets, even from across the road? Or is there something else here than meets the eye?

Keep in mind how much that must be costing others in ways hard to quantify: engineers around the world struggling tirelessly to debug essentially jinxed boards, faulty products released to the market, expensive recalls, maybe even some promising careers jeopardized, and so on. However, to be fair to the company's engineers who were originally involved in pushing out this junk IC family, or at least by sheer statistics, one or two of them may actually have spotted the problem at some point over the years. Maybe their seniors were just not around long enough to guide them properly (very likely if you ask me). Or maybe the engineers were just given a big "shush." But it is also possible the engineers just relegated it to being a bad part, trashed it, and moved on.

And therein lies an important lesson for us, too. We must never ignore a problem. Never assume anything. We need to test and analyze not only parts that work, but even those that don't when there is no obvious reason. We must send every such bad part for further failure analysis until we are sure it really is a non-chargeable failure. For example, if in some case we can prove the PCB/layout is the culprit, and then we go ahead and fix it, that specific failure mode is considered non-chargeable thereafter. It is only then that we can truly afford to forget about it. But till that time, by definition, the failure is still chargeable, and investigate it we must. For all you know, that bad part could turn out to be the only valuable advance information of an impending and devastating recall that could have been averted in time.

Know Your Instruments Well

Sometimes the so-called problem could just be an artifact of the instruments being used to characterize or test the converter. That can be really embarrassing to find out, especially after alerting everybody from Design to Production! Backtracking too many times, especially in the course of a single day, can become a rather overwhelming declaration of incompetence. So it is extremely important we understand our instrumentation well. Because if we don't, a more serious situation can arise—we may fail to capture an existing or incipient problem until it shows up at the *customer's* doorstep. And we know all too well that customers have recently developed this almost uncanny penchant for catching whatever we missed. Yes, we could argue that we now form a great team together! The only question is, for how long?

Heatseekers: On the Road with Symptomatic Troubleshooting

All of us have, sooner or later, had to sit rather nervously in a physician's waiting room, with the staff buzzing around looking extremely grim-faced at us for some strange reason. I have actually had that experience in more than one country. Which is how I think I gained some additional insight into the different ways doctors across the world handle their patients. In Germany, for example, you can die a thousand deaths right in front of their eyes, but they may never give you even a simple painkiller till they have a *Korrekte Diagnose* on hand (countersigned by another doctor, mind you). Maybe that's why they call their outpatient department an *ambulanz*. The complete opposite of that is in India, where doctors often dole out a bag of gratifyingly multicolored super-antibiotics right off the bat. In fact I remember at times, I was not even done fully describing my symptoms and the doctor had finished scrawling out the full list of medications (and even called out "Next!"). This is what I call pure, inspired symptomatic troubleshooting (whereas the first method is probably best described simply as German, for want of better words.) Actually, both methods do work on occasion, but you need a lot of luck to get the symptomatic method to work in the long run. The main problem with it being that it can also prove fatal on occasion!

Putting all this in the context of power supply design and troubleshooting, the following episode should more clearly illustrate the perils (and signs) of symptomatic troubleshooting.

You may be aware that electronic ballasts, as used, for example, in fluorescent tube light fixtures, are also based on switching power conversion principles, the main difference being the energy delivered is finally light. This story goes back many years, when I was a struggling engineer in the central R&D lab of a major electrical manufacturer in Bombay (think of it as an Indian Siemens with a finger in almost every electrical pie in the works, and likewise, almost always losing money in the process). I had somehow gotten assigned to finally solve all the problems still haunting their long drawn out ballast development. The original team, working out of the huge lighting division several miles away, had apparently been in the midst of a heroic saga for the last two years, having acquired some notable skills in creative reverse-engineering along the way. Dr. G. T. "Doc" Murthy, my vice president at the time and head of R&D, incidentally a brilliant ex-physicist from MIT (yes, the one in Massachusetts), was taking an unusually keen interest in the project. As it turned out, for the first month I was simply going to be his apprentice. It was obvious he really wanted to be the one who solved the problem. Of course, Doc's unusually high interest might have had something to do with the fact that the CEO of the company had set up his own (privately owned) mini-company on the outskirts of Bombay, with the sole mandate of manufacturing (in astronomical volumes) the ballast being developed by us. The last leg of the plan was to sell the ballast back to the company he ran. And which purchase officer sitting there would not agree it was the cheapest and most reliable ballast ever made! Unfortunately, customers would probably have a mind of their own. So yes, margins were crucial, and they also knew they couldn't afford massive recalls or rejections. Clearly, with so much at stake (for so few), you could be reasonably sure that out-of-turn promotions awaited the truly deserving. The only stumbling block was that damn pair of overheated, unreliable switching transistors of the Half-Bridge inside the ballast.

For the next month, Doc would suddenly materialize every evening at my bench, brandishing a sheet of freshly hand-scrawled calculations. After some eloquent hand waving and a cursory explanation, he would ask me to quickly build his fascinating new idea. An hour or two later he would reappear, impatiently waiting for me to power it up, exclaiming, "watch, now the transistors will run real cool." So I would turn the ballast on, run it for about five minutes, then unplug it from the 240VAC mains (to avoid electrocuting my beloved mentor of course!). At which point Doc would lean over excitedly to touch the exposed transistor heatsink. Now, if you have ever put your hand in a lion's cage and got it completely bitten off, you would realize how Doc probably felt in the moments that followed. Luckily, the building was almost vacant by the time his scream hit the walls. Good thing too, because that's the only way we could continue to put on a brave face during normal office hours for weeks thereafter. It became our own private little secret!

But finally, one evening I chanced upon Doc glowering at the ballast, saying, "A handful of components and it attacks me every time like a tiger!" With that he stormed off for good, leaving me effectively in control of the project for the next month. That's how I somehow managed to be able to solve all its problems, and also managed to reduce its BOM (Bill of Materials) cost by almost half. I guess I had been lucky.

Rather than tell you how I heroically and almost single-handedly knocked down the price, I will describe why the cost had spiraled up so high in the first place, and why the reliability had taken a sound beating. Because that was actually the real problem here. Take a look at Figure 1-5, and try to uncover a lesson in engineering embedded within it somewhere. We are not trying to be too technical here. This figure apparently describes what the previous team had been doing all along. It is a stark reminder of the pitfalls of symptomatic troubleshooting (i.e., hitting out blindly at symptoms, not their causes). Initially, they had experimented with the base drive of the transistors, and that helped lower the operating

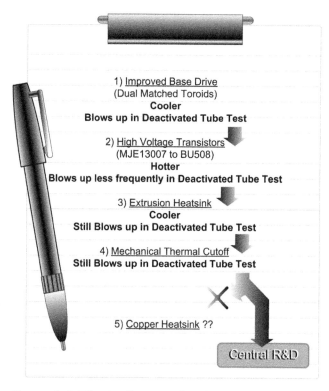

Figure 1-5 The Perilous Path of Symptomatic Troubleshooting

temperatures. However, their solution was expensive, requiring painful matching of magnetic permeabilities on the production floor. Thereafter, to get the transistors to survive a certain critical abnormal test, they upgraded to much higher voltage devices ("let's just get a beefier transistor"). Probably unforeseen by them was that higher voltage devices also have higher forward voltage drops and longer transition times. Therefore both the switching losses and the conduction losses escalated steeply with this step, along with the operating temperatures. To solve this new emerging problem, they brought in an expensive aluminum extrusion heatsink. They probably realized it would do nothing to improve the efficiency, but it seemed worth a try. However, the transistors continued to blow up at an alarming rate in the abnormal test. So in came an expensive mechanical thermal cutoff, mounted a few centimeters away from each transistor (midway between them) and screwed on to the heatsink. This didn't help either. Because, under a sudden abnormal condition, the power transistors would always heat up very suddenly, and the heatsink clearly had a hard time keeping up. A steep temperature gradient was being established in the moments that followed, and the junction temperatures of the transistors no longer bore any simple or predictable relationship to the temperature of the surrounding heatsink. And that's apparently when the call to 911 (Central R&D) went out. Though I believe the President of the company himself had to get involved, to make the somewhat recalcitrant ("can-do," but "*when*-do") engineers finally admit they desperately needed help.

Many years later in Singapore, we were using a specially formulated thermally conductive glue to fix the overtemperature sensing thermistor smack on to the very plastic body of the TO-220 power transistor. We had empirically ascertained that in this way, the junction temperature and the adjacent temperature as seen by the thermistor were less than 10°C apart, even during an abnormal event. So if, for example, we wanted to have the transistor turned off just before it hit 150°C, we simply needed to set the trip temperature (of the thermistor-based circuit) at about 140°C. In that way, we could also be sure that we wouldn't encounter nuisance tripping on a particularly hot day, when the temperature inside the enclosure would also be much higher.

As an aside, keep in mind that anything mechanical is usually going to cost a lot more, such as switches, relays, connectors, heatsinks, and so on. Mechanical devices can also be relatively unreliable in the long run, so try to avoid them if possible. They are also harder to debug because they can often end up creating *intermittent* failures—the hardest category to catch and analyze. For example, don't ever troubleshoot or design switching power supplies with trimpots (small rotary potentiometers) present! Instead, take the extra trouble of soldering fixed resistors at each step, if necessary paralleling them if you need to vary the resistance even slightly. Also, prefer thermistor-based electronic latches to mechanical thermal cutoffs, and so on. Don't forget the PCB itself is a giant mechanical device, one that you certainly can't do without. But at least get it made from a quality vendor. However, even with that precaution, I have seen many cases where some buried via was just not connected properly for some odd reason. So, if you flex the board slightly, it makes contact and the converter comes to life. Leave it, and it's as good as a disable pin. What you can do to avoid such inescapable defects is to create some form of redundancy. It doesn't always cost you money. So I learned to always put in several vias in parallel, even where one may have sufficed. Not only does that lower the inductance, but if one via is defective, the others would take over the job and no one would be the wiser for it.

We should also remember that every protection circuit, electronic or mechanical, can under certain conditions become a nuisance (and danger) on its own. Which is why, when starting to troubleshoot, I usually first try to carefully deactivate any ancillary circuits such as overcurrent trips, overvoltage protections, and so on. That way I can confirm the behavior of the bare-bones engine, the *switching power stage*. Only if I confirm that that is indeed trouble free, do I start reattaching the auxiliary circuits, block by block, until the problem suddenly shows up. And thus I can immediately identify which specific circuit block is the culprit and dive into that more deeply. The process is akin to taking a film of yourself peeling an onion, then playing it back in reverse (reverse-peeling). This can be a particularly effective debugging approach, especially when dealing with more complex power supplies (i.e., AC-DC types).

Returning to the case of the ballast, because the heatsink was a huge aluminum extrusion (with a correspondingly large thermal capacity), it actually took a significantly longer time to heat up once it was added. So any knowledge of the rapidly escalating transistor temperatures arrived at the thermal cutoff too late. And by then, the ballast would usually be Sunday morning toast. It was probably mystifying to the engineers that even by lowering the steady operating temperatures (by means of the heatsink), they were actually incurring higher failures in the abnormal test (despite their new triple diffused 1500V horizontal deflection transistors!). I suspect that if they had not been instructed to get help at this point, they would have been trying out thick copper heatsinks next, to get the thermal cutoff to react more quickly.

We can see a distressing, but not too unfamiliar a pattern emerging here, one wrong move leading to another, to another, and so on. In wise-guy circles this is sometimes known as the Nixon principle, "if two wrongs don't make it right, try three." In power, it is called symptomatic troubleshooting.

We can never hope to converge to a valid and optimum solution if we are unwilling to do what was initially suggested: develop an ability to look at any practical problem through the eyes of its supporting theory and vice versa. I think that's how I eventually succeeded. First I created a rather textbook base drive circuit (not expensive). But to do that I had to go back to first principles and educate myself further on what is the best way to drive NPN power transistors, the exact waveshape required, and so on. Then I created a "balun drive," which was, I still think, a rather clever trick to produce symmetrical drive waveforms for both transistors (to avoid cross-conduction). The balun drive would also guarantee that symmetry was maintained over all ferrite and transistor production spreads, without having to screen and match the magnetic base-drive cores of the two transistors anymore. All put together, that directly solved the steady state temperature problem completely. The efficiency had also improved significantly, because I had actually lowered temperatures by reducing the losses, not merely using bigger heatsinks to get the heat out. And I could also see I had not made the second problem (failures under abnormals) any worse. The solution to the first problem had truly converged and was therefore valid.

After the above steps, I carefully tracked down the real cause for the second problem—the transistor blowing up in the abnormal test. That actually called for a lot of eyeballing of various oscilloscope waveforms (the investigation phase). I also put in some more reading and plenty of thinking (the analysis phase). Finally I realized it only called for a thin extra winding placed on the choke attached to a small cutoff circuit consisting of two tiny transistors (forming a classic NPN-PNP latch). That solved the second problem, and very cheaply too. I had to confirm that neither solution ended up having a secondary impact on any other aspects of the performance. So for the second solution, I had to carefully calculate the exact number of turns of the extra (energy-recovery) winding.

If interested, the lurid technical details of the ballast solutions discussed previously are documented in the Appendix of my previous book *Switching Power Supplies A to Z*. In this chapter, I only wanted to focus on the lessons learned, and these have therefore been presented as such with all their supporting cast and props!

In retrospect, what saved the day for me was my willingness to dig deep into theory, fully understand the root cause, and then try to solve it by putting pen to paper. So if you usually find yourself in the lab without a calculator or even pad of paper, you are probably heading down the well-trodden path of symptomatic troubleshooting toward certain disaster.

Causality Can Be the First Casualty

We can extract several important lessons from episodes such as the one described above. Let us list some of them as follows:

1. Try to mentally separate problems. Just because they seem similar does not mean they are the same. So a transistor that *heats up and blows up* is not necessarily the same problem as a *transistor that blows up when it is hot.*

2. Of course, later you should see whether the problems correlate somehow. Sometimes a single cause could be creating *multiple symptoms*. For example, a bad PCB layout may create poor output regulation and also device failures.

3. Similarly, a single symptom could have *multiple causes* too. For example, a switch failure during power-up or power-down could be the result of an inadequate transformer design and inadequate duty cycle limiting in the control circuit combining forces inside your AC-DC Flyback. There are actually two things to fix here, not one. Take another example of a shoot-through failure occurring in your Synchronous Buck, which may be the result of both inadequate deadtime and poor Fet selection. Looking at more complex problems such as this one: a device failure on your current mode control Non-Synchronous Buck during output shorts may be the result of a very high switching frequency combining forces with too high a minimum on-time, and too low a diode forward drop, causing the fault current to freewheel without slewing down sufficiently during the off-times, thus causing uncontrolled current staircasing, without the current limit being able to do anything about it. Quite a mouthful, but see Figure 1-6 to understand this a little better.

4. One of the hardest to analyze and fix are probably the *chain reaction* problems. For example, in an AC-DC power supply, we know all too well that the switch can blow

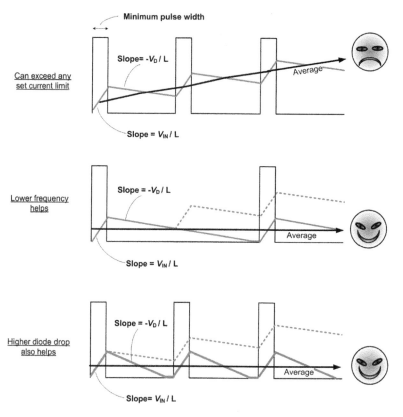

Figure 1-6 Unrestrained Current Ratcheting, and Ways to Prevent It

up under certain conditions. Moments later, the sense resistor in the Source of the Fet fails to open due to the excessive current. This diverts the huge surge of current from the Drain into the Gate of the melting Fet, and then onto the control IC. After damaging the control IC by blowing up its ground pin bonding wires, the current (continuing to be pushed through by the inductance) finally freewheels into the opto-coupler connected several inches away, via the intervening PCB traces. To the observer, the switch and the opto-coupler seem to have exploded simultaneously, several inches apart. Maybe something on the secondary side also blew up next, such as the TL431. See Figure 1-7. At first, you should resist the thought that the opto-coupler may have failed first, causing the switch to fail. It is more likely the other way around in this case. But neither should you, in principle, ever overlook the slim possibility that the events proceeded in exactly the *reverse* direction to what you instinctively tend to believe. It is always important to keep an open mind. So, just as an example, suppose the TL431 exhibited a strange oscillation, and that was what caused the switch to blow up, leading to the chain of events described above, and thereafter, also caused the destruction of the TL431 itself. The cause and effect have thus finally converged—the invention basically returning to plague the inventor. (I, for one, am always doing that in real life too—initiating a chain of

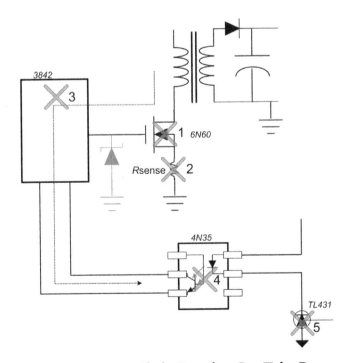

Figure 1-7 How a Chain Reaction Can Take Down the Opto-coupler Several Inches Away

events that comes around full circle and bites me in my pants!) This opto failure incident actually really happened to us while seeking UL approvals several years ago. We learned that the cheapest way to divert the current away from the IC (and also prevent plenty of rework to get the power supply to work again) is that small 18V grayed-out zener in the Gate of the Fet in Figure 1-7. The zener fails shorted when the switch fails, and stays that way until the fuse blows and the inductor energy is spent. That way the safety barrier (the opto in this case) is not breached during UL testing. In fact, that is the only reason we finally put in a Gate zener into our commercial models. It was always a matter of cost for us, and we needed a very good reason to add even a cent to our BOM. However, while troubleshooting AC-DC power supplies, I now always put in this zener on my initial prototypes (I plan for it at the PCB design stage). After any blow-up (and that's three times a day on an average!), you rarely ever need to replace anything other than the input fuse, the Fet (of course), the sense resistor, and the zener.

5. Another possibility to consider is that of *hidden modes* of operation. I would say these are truly the hardest to figure out, simply because you didn't know that particular mode even existed prior to that. This may or may not be a flaw in the control IC. Maybe that is just the way it is. For example, once, we were seeing some sporadic, utterly mysterious failures of the Fet in our 3842-based Flyback whenever 270VAC was suddenly applied at its input. After a lot of investigation and analysis, I finally had my theory validated. It was happening like this. When power was suddenly applied, there was naturally a sudden increase in the voltage at the Drain terminal of the Fet. But some of this rising dV/dt fed a current through the parasitic Drain-to-Gate capacitance, pulling up the gate. See Figure 1-8. Though the Gate had a hardwired 10kΩ pull-down resistor to ground, it was apparently not

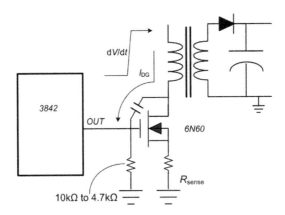

Figure 1-8 Spurious Turn-on of Fet under Power-up

enough, and there was in fact a huge bump visible at the gate. The reason the 3842 was not able to prevent the catastrophic failure that followed was that it still had not powered up fully. And we then learned that if this IC has not yet come up fully, its Gate drive pin (labeled OUT) is in effect tri-stated, that is, *floating*. That was a hidden mode of this popular control IC, one that we were not aware of till then. I think we just hadn't thought about it. Ultimately, to solve the problem, we had to improve the pull-down at the Gate of the Fet. Its value was changed from 10kΩ to 4.7kΩ on all future models. All failures stopped and we knew that from then on, we would always *expect the unexpected*!

To "Errr" Is Human

There is another practical issue to circumvent too. Remember how many times we have scribbled out some very seemingly impressive calculations (and often even built something based on them!), only to discover they were embarrassingly wrong to start with? And that could have happened for any of the following equally impressive reasons; a) the handwriting was almost illegible, b) the formula was missing a factor of $\sqrt{2}$, c) the units were wrong, d) the math manipulation went awry when we transferred the terms to the other side of the equality sign, or e) you wrote 20% tolerance where you actually meant ±20%, that is, a 40% tolerance band.

In 2005 it became grindingly obvious that there was a certain spot on Seattle's monorail where tram cars trying to simply pass each other by had to stop to take samples of exterior paint from each other. That was traced back to an engineering error they had made back in 1988, when the rails were brought in a little too close! Somebody get me a ruler please!

I read recently that in 1979, five nuclear reactors in the US were shut down temporarily, because a program testing their resistance to earthquakes used an arithmetic sum of variables instead of the square root of the sum of the squares of the variables. Actually, I have done that many times myself while calculating the RMS (root mean square) of a waveform!

In 1985, the Strategic Defense Initiative Organization performed a simple experiment (and thereby rose to the occasion). The crew of the space shuttle was to position the shuttle so that a mirror mounted on its side could reflect a laser beamed from the top of a mountain 10,023 feet above sea level. But the computer program controlling the shuttle's movements interpreted the information it received on the laser's location as indicating the elevation in nautical miles instead of feet. As a result the program positioned the shuttle to receive a beam from a nonexistent mountain 10,023 nautical miles above sea level.

In 1999, a $125 million Mars space probe crashed mysteriously, only because it later turned out they had gotten British and metric units all mixed up in the software!

It is interesting to follow the last case mentioned above a little more closely. I started wondering—did they finally smarten up, but just enough to fire all the errant engineers involved (heads on the platter my lord)? Did they throw all their "perfidious computers" off the roof of Building D ("I knew it! Gates is behind everything that goes wrong nowadays")? Luckily not! Because, as a certain key figure of NASA later put it quite aptly (the italics are mine):

> Sometimes people [do] make errors . . . [but] the problem here *was not the error* [itself].
> It was the failure of NASA's systems engineering, and the checks and balances in our
> processes to *detect* the error. [Italics added—SM]

(By this logic I can finally start blaming all the heartrending typos of my three books on the publishers!)

As a corollary, if you work in a "trusting" atmosphere with great teamwork, you automatically tend to wholeheartedly believe the data presented to you by a colleague, and probably feel very guilty (or even sneaky) verifying even a single line of it. But my question to you is, where are the checks and balances? As I wrote in the Preface: people who help you find mistakes before you make them are actually your best friends in the business. So help them help you. Don't act defensive when someone tells you how you screwed up. Thank them, because at least now you know you won't be that stupid ever again.

At least on your part, errors can be reduced if you learn to automate (wherever it makes sense of course). That will also leave you mentally unfettered to really think about the problem. Use computers, but like any other tool out there, use them wisely (and correctly). They are not the real problem. In fact you may be the last remaining problem, for not realizing or admitting that you need to learn how to use them. The good news is that once you have gone through the initial trouble of creating your very own validated (cross-checked and peer-reviewed) spreadsheets, life is much easier (and more accurate!) thereafter. You can churn out repetitive iterations at the press of a button. Now at last, you don't have to redo every calculation every single time, making a proportional number of mistakes in the bargain. You may prefer to use Matlab, Excel, or something else. My tool of choice is Mathcad (but that may simply be because that's all I ever learned). It does serve my purpose well. I have found it easy to prepare a universally readable report from it, prepare an easy tutorial for my Apps (applications) team, and even send it off to the customer in text format if required.

For similar human reasons, we should document every single oscilloscope plot carefully the moment we capture it (assuming it seems meaningful of course). Write down the input and output voltages, currents, specific applied conditions (i.e., power-up into short circuit), the state of the other pins, and so on. Don't forget to keep close track of what each channel represented (or later, just watch yourself suffer: "Hey look, by moving the pole-zero pair apart, I now have negligible overshoot at start-up. . . . Oops, that must have been the Enable

Pin waveform, not the output"). Also record whether the waveform is a one-off event or a repetitive occurrence. Also, you may want to number and log the specific unit (IC) involved to send it for failure analysis if necessary. Record anything different about the PCB being used at the time, such as, did it have a current probe hooked up to it that may have been causing or aggravating the problem? Later, when you sit back in your sunny little cubicle to really analyze the data, your conclusions are going to be only as good as your hazy memory. After barely a few days, none of the scope plots will make much sense to you anymore. The harsh reality is that as you acquire more and more experience to be better able to analyze the emerging data, ironically, the more likely you are going to forget how it ever even came to be (that's a process they call aging). How often have I gawked at some old thermal paper scope printout on my table, wondering who in the world managed to forge my handwriting so convincingly!

Learn from (Their) Mistakes

Keep in mind that the ultimate idea is to always try and learn from mistakes, yours, mine, ours, and theirs (preferably the latter!). We stay alive by constantly questioning our own rationale and assumptions. The worst favor you can do to yourself is becoming smug as a bug ("hey, you can call me *Guru* from now on"). That is the surest recipe for disaster in the long run. If you don't believe me, try loudly enunciating your latest pet theory to a peer, and you may suddenly realize how unconvincing it really sounds (even to you). How you stumble over the most obvious steps to the sound of your moving pleas for a leap of faith every now and then. It is just not as good as you thought it was! Unless of course, you are a real guru! Then hats off to you. And I really mean an *engineer's* hat of course.

In power conversion, the most obvious conclusion is often the wrong one. At best it may simply not apply. But more often, if we think real hard, we realize it calls for a sea change in the basic laws of Physics. And that's usually when we sheepishly slink out of the phase review meeting, mumbling, "I have a quick errand to run, be right back." Surprisingly, we all have been there at some time or the other.

But if a decade later, you still haven't learned from even your own mistakes, that may not be graciously accepted as human error anymore. Humans are also supposed to fix errors, however human they may be! For example, in 2007, it surfaced that human error was behind the 2006 cascade of events that caused all sorts of problems on the $154 million Mars Global Surveyor, leading to its early demise. According to NASA, its engineers while doing a routine update sent a stream of incorrect software commands repeatedly, then "did not catch their mistakes because the existing procedures to do so were inadequate. . . . Had these procedures been more rigorous . . . then perhaps this wouldn't have happened," said their board chairperson. She only forgot to add "Déjà vu!"

The Problem with Problem Solving

As I mentioned earlier, don't ever hesitate to Google shamelessly when you're stuck. You may get extremely quick answers to your problems based on others' shared experiences. The web has truly accelerated the pace of our learning, and the good thing is most engineers love to share. No longer do you have to depend on somebody to walk past you near the coffee machine, offering some meaningful, off-the-cuff advice. All you need now is the right attitude.

And that's how I discovered the following interesting piece by Ronald L. Hughes, Senior RCI Consultant & Trainer, originally published in *Plant Engineering Magazine*, January 2003. Except for the **bold face italic** content, which I have added, I will quote him verbatim (because he says it best):

> *Good problem solvers are among the most highly sought after individuals in existence today. In most cases companies and organizations have already identified their skilled problem solvers and consider these individuals to be amongst their most valuable assets. Their expertise is recognized and their time allocated for solving the most important outstanding problems. With this being the case the question becomes, why do so many problems still exist? The answer becomes readily apparent when we realize that the science of problem solving often takes a back seat to the other more pressing day-to-day issues we face during everyday operations. These perceived time constraints rear their ugly heads under many different guises and are the main reason why true problem solving does not receive the effort or the support it so rightfully deserves. The prevalent paradigm being—"we don't have time to analyze a problem—just fix it." I contend that **if we indeed have time to fix a problem over and over again, and for the same apparent reason, then we have the time to analyze why the problem is occurring** and take the proper measures to eliminate or significantly mitigate the consequences of that problem.*
>
> *The real problem is we just don't have enough true problem solvers to go around. With ever increasing demand for the problem solvers' time, true problem solving is allocated fewer and fewer resources and thus returns less and less benefit to the company. This phenomenon is worth further exploration.*
>
> a) **Short-cut Problem Solving Techniques**
>
> *As managers, it is incumbent upon us to finally admit that we want the answers to our outstanding problems right now! We want the problem to go away immediately so we can deal with the more pressing issues of our everyday responsibilities. As managers we must also realize what we are telling our subordinates when we have a "fix it now—analyze later" attitude. The obvious conclusion derived from our actions is that there is no time for problem solving in any form, in short, problem solving is just not allowed. This attitude permeates throughout the organization and promotes short-cut problem solving techniques like part(s) replacement. The technician feels that he or she must get the equipment or process back up as soon as possible, taking any and all*

short-cuts available. The best way for the technician to accomplish this is to start changing parts (based upon passed experience/success with correcting the problem) until the equipment or process starts producing again. This is certainly not the message we should be promoting and is definitely not the way we should be using our highly skilled technicians. Not only is it counterproductive to our staff, but to our operations and the resulting bottom-line as well.

b) **Disconnected Problem Solving Techniques**

*Due to a lack of understanding in the use of analytical skills, most people resort to **disconnected problem solving techniques** to analyze their problems in lieu of a structured logical approach. This stems from the fact that we do not give analysts the tools necessary to do their jobs. Simply put, many of today's analysts lack the proper mentoring and training necessary to accomplish the desired result—the elimination of problems. Without these tools these analysts revert to their inherent god-given analytical techniques; i.e., inference, perceptions, assumptions, intuition and reports by others.*

c) **Inference**

Inferences are decisions individuals make based on their own personal logic systems. For example, an excess of 90% of all bearing failures have some sort of fatigue mechanism associated with the failure. Therefore, the tendency is to associate all bearing failures with some form of thermal or mechanical fatigue. The problem with this is that if the bearing keeps failing over-and-over again, the analyst may be looking for a cause that does not exist—by searching for fatigue based on their inference, thus overlooking the other possibilities that could be associated with the failure—overload, corrosion or erosion.

d) **Perceptions**

*Perceptions are what our five senses tell us. The problem with perceptions is that they often will fool us into drawing a conclusion that is not factual. For example, **if you think you see a fire your perceptions trick you into smelling smoke**. A case in point, light refractions are often misinterpreted as flames flickering, thus smoke is smelt, and heat is even sometimes felt by the observer. Here it is important to understand that we don't always see what we thought we saw, hear what we thought we heard, etc. Dealing strictly with facts will eliminate the misinterpretations of perceptions. In our example, it would be easy to verify that a fire was actually burning versus an electrical short that was arcing, or the simple refraction of light from a luminous surface by going to the location of the incident and visually inspecting the area in question.*

e) **Assumptions**

Assumptions hinder the problem solver because they tend to apply a rule or regulation to the problem that does not exist. For example, by applying the correct torque to a piping flange setting, the assumption is that the clamping force of the bolt will ensure a good seal between the flange faces and the gasket material during normal operations. What is missed here is the fact that it is not the amount of torque you put on the bolt, but the amount of tension that is put in the bolt that determines its clamping force. The assumption completely overlooks the condition of the fastener's threads at the time of assembly (pre-load) and the fact that once the system is heated up, some of the fasteners will have increased tension in the flange setting, while others will have reduced thread stretch due to thermal growth during startup and operations (final load). When analyzing the example, look at all the assumptions that are made. The flange faces are perfectly square and aligned and will remain so throughout operations, the mechanical fastener threads are in good shape and will act as perfect springs during assembly, there is a perfect torque/tension relationship in each fastener set, the operational conditions of the system will not affect the tension of the bolts in the flange assembly, etc.

f) **Intuition**

*Intuition is defined as our "gut feel." Studies have shown that our intuition does indeed serve us well as we are intuitively correct approximately 70% of the time. The problem here is that **we are wrong approximately 30% of the time** and our problems still continue to remain. This should be a signal that our intuition is steering us in the wrong direction. However, without the proper training in the use analytical skills, the analyst continues to rely on his/her own internal logic system—intuition—as the tool to be most likely used when analyzing recurring problems.*

g) **Reports by Others**

*We all know individuals in whom we place a large amount of confidence or credibility. We rely on these people to give their honest opinions and recommendations when asked about difficult problems. It is common practice for us to seek out their advice and then act upon their recommendations based on past successful experiences. There is nothing wrong with doing so as long as we remember that what they tell us often employs the same analytical tools that have just been discussed; i.e., inferences, perceptions, assumptions and intuition. In short, **they also could be wrong**.*

So how do we keep disconnected problem solving techniques from driving our analysis efforts? The solution lies in dealing strictly with factual information instead of any other analytical tool.

h) **Fact Based Analysis**

*Fact based analysis—the elimination of "what if" scenarios. **Start with fact, end with fact, and what you have is fact, not supposition**. The process sounds simple enough but is seldom used. The key is to analyze using short deductive steps in logic, and then verifying at every step during the logic development process. By taking these short deductive logic steps, all of the logic holes are covered. Also, by verifying at every step, the validity of the analysis is self-evident. There is a great amount of confidence* *in the analysis when all the possibilities have been explored and when the validity of all hypothesis verifications is proven sound.*

*Probably the most interesting and enlightening analysis you could ever perform is to analyze why failure analysis fails. By its pure definition, **it is impossible for failure analysis to fail—so how come it does? The answers (root causes) often lie with the problem solvers themselves**, as well as their management support groups.*

i)   **Problem Solvers**

*Quite often, the problem with problem solving is the problem solvers themselves. Without the proper training analysts tend to solve failures by going straight from their event to a cause using one of the disconnected problem solving techniques previously discussed. This is the way individuals think—in a straight line. What they fail to realize is that **problems seldom occur in a straight-line pattern**, or for a singular reason. What is missed by not using short, **deductive steps in logic**, is the analysis portion of problem solving.*

j)   **Order and Pattern**

*There is order and pattern to everything in the universe. Likewise, **there is order and pattern to failure**. The key is how to uncover the order and pattern of the failure. First one must realize that the failure is actually being looked at in reverse. The root causes are actually the point at which failure began and the event is merely the result of the root causes or how the failure manifested itself. Second, there is a direct cause and effect relationship that can be associated with the order and pattern of the failure when analyzing. Finally, the cause(s) always go below the effect in any fault-tree type of analysis. For example, **does misalignment cause high vibration or does high vibration cause misalignment?** The answer to both questions is "yes." **So how does one know which is the cause and which is the effect?** Here it is good to understand that there are various determinable results from any input. Using our misalignment example, when analyzing we must also consider that the equipment was either initially misaligned or the equipment was aligned correctly and became misaligned. Understanding this simple principle it becomes easy to determine if high vibration caused misalignment (became misaligned) or misalignment caused high vibration (initially misaligned). Once the cause has been determined then it becomes the effect and subsequent causes of the newly identified effect are explored. This process—the cause and effect relationship, is reiterated until all the roots have been uncovered, or the order and pattern that led to the failure has been determined.*

k)   **Data**

*Problem solvers do not always understand the importance of data to analysis efforts and are therefore poor at the identification, collection and use of failure data. **Data is definitely the key to successful analysis**. With every piece of data obtained, the analysts should be asking themselves questions relative to what the evidence is providing. For* *example, if you clean up four quarts of oil after a failure on a piece of equipment that should have contained six quarts of oil the obvious question is "where are the other two quarts?" This type of deduction literally is **building the logic tree** for the analyst. Here the key is to understand that the evidence provides the answers. **Let the data push the analysis to successful conclusion**. What the evidence tells you is fact and what we are interested in uncovering is the facts of the failure.*

Here it is good to note that data not only is essential for successful analysis, it also determines the speed of the analysis effort. In essence, the more data you have prior to analyzing, the faster the analysis will go.

l) **Management Support**

*Another key factor, or root cause, in the unsuccessful results of failure analysis is the lack of management support for problem solving efforts. This stems from the fact that problem solving is thought of as an effort that we expect from all employees within the organization. What is not understood is that we don't provide our employees with the necessary tools to do the job. True problem solving only takes place when the proper support mechanisms are put into place. This is the job of management—to identify and provide the necessary tools [**including people—SM**] to make problem solving actually work and provide the desired results.*

m) **Culture Change**

For problem solving to provide the maximum return, a change in culture is definitely required. Many of us say that we support problem-solving efforts, few demonstrate what they are saying by actually implementing polices and procedures that not only support problem solving, but demand results. It is not enough to just do problem solving, there must be a means put into place for tracking the results of our efforts. Problem solving should not be merely accepted, but expected, with pre-established returns on investment set at extremely optimistic levels.

n) **Training and Mentoring**

*It is truly amazing how ineffective our training efforts have become. Taking a closer look at the phenomena provides us the cause(s) of the effect. We send our employees to training and expect them to sit in a seat for 1 to 5 days and magically absorb the information presented without actually giving them the opportunity to practice what they have learned. Training, as an end to itself, is not a cure for the problem of ineffective problem solving. Again we must expect a return for our investment by requiring the employee to demonstrate their newly acquired knowledge by actually performing the skill they have been taught. **This should be done immediately after the completion of training—before the knowledge is lost**. Studies by the American Society of Training and Development have shown that if this is not done within the first 3 to 5 days following the training session, then what was learned is either significantly lost or will not be used at all.*

*Once the employee has demonstrated the basic skills for problem solving it is then management's role to perfect that skill by providing a mentoring program that will further hone the capabilities of the new analyst. In this case **providing access to a seasoned expert** in the science of problem solving to the new analyst. The expert's job is to mentor them through their failure analysis efforts. Remember what a mentor actually is—one who has the required knowledge and provides it to those who don't. Here it is management's job to make sure that this happens. Unfortunately, not all organizations have seasoned problem solvers on their staff. If this is the case, they should be obtained from outside organizations specializing in this type of work. The key here is to recognize that although we have many qualified employees who do their individual jobs well, this does not necessarily make them experts in the science of problem solving.*

(Reprinted with permission from Ronald L. Hughes, 2003, The Problem with Problem Solving, Plant Engineering Magazine.)

By now we have hopefully acquired the mindset needed for successfully troubleshooting switching power supplies. It is now time to start diving deep into technical details, because that is where the devil really lies. Power supplies are nothing if not all about seemingly minor details. Though that fact itself is a detail a lot of people miss, especially from 30,000 feet in the air!

High-Frequency Effects and the Importance of Input Decoupling

Lies, Damn Lies, and Schematics

A schematic hides more than it reveals. For example, it seems to imply that that every point of a circuit with a ground symbol hanging from it happens to be the same point (i.e., at the same voltage). Nothing could be further from the truth, especially when the PCB is faced with the job of transporting the high-frequency current harmonics associated with switching transitions. That is why it is never enough to simply look at an electrical schematic and then build or troubleshoot a switching regulator.

PCB Trace Impedances

A few millimeters of PCB trace length can become a veritable impedance wall for such harmonics, causing the voltage at one end of the trace to lift up (with respect to the other end), producing an unintended result somewhere or other. And when that impedance is predominantly inductive, this voltage kick can be really nasty, based on the simple equation $V = L dI/dt$. Here dI/dt is the slope of the current edge, and L is the inductance associated with the trace section. Let's do the math here. A typical DC-DC converter may switch several amperes in 20ns. The rule of thumb for PCB trace inductance is 20nH per inch. So if we switch 1A through 1 in. of trace, we will get 1V of kickback. A *switch* of 2A will give 2V and so on. Depending upon where the offending PCB section is located, we could get voltage spikes being applied to the pins of the IC. This could affect pin thresholds and cause erratic behavior. For example, if the IC has a clock pin, this could produce severe jitter in the switching waveform, which in turn could produce other measurable effects. The good thing is that these inductive spikes usually do not have much associated energy, and therefore tend to get absorbed partially by real or parasitic capacitances nearby and/or get dissipated in adjacent resistances. But for *dissipation* to really occur, the *voltage* spike has to be able to drive some *current* (in a closed path). Unfortunately, if that current path inadvertently includes unspecified circuit blocks inside the IC (ESD structures, substrate, etc.), we could not only cause erratic switching but general controller malfunction of a temporary or permanent nature. That is why the topic of PCB design acquires such importance while designing and troubleshooting switching regulators.

Ground Bounce Inside an IC

Modern power IC designers have started trying to fight ground bounce inside their chips by bonding out separate pins labeled AGND (for analog ground), DGND (for digital ground), or PGND (for power ground). The purpose is that if, for example, the digital block suddenly draws some current, the sudden inductive drop across the bond wires does not cause any momentary (but likely momentous) internal ground reference imbalances, resulting in false communication at the various analog/digital/power section interfaces inside the IC. Note that these separate ground pins are usually meant to be connected very closely together on the PCB, onto a copper island, and AC-decoupled well with respect to the supply pins (more on decoupling later).

That reduces the ground bounce caused by the internal bond wires of the IC, but getting the entire converter to work properly requires a deep understanding of the role played by the PCB, *its* contribution to ground bounce, noise, and so on.

The Ground Plane

If we look at the PCB of a real converter, we don't really know where to even start demarcating noisy areas from quieter ones (on what is supposedly the same ground). So usually we hope that by just throwing in a thick underlying copper layer (the ground plane, a dedicated layer of the PCB) and *stitching* it at fairly closely spaced intervals by means of vias to the various ground nodes, ground terminals, and ground traces on the component side, we will survive. The good news is we usually do! But eventually, a lot depends on how good the ground plane itself really is (vis-à-vis our specific application), and how effective it therefore is in helping us equalize all the grounds of our circuit. Of course we will never manage to fully equalize the ground nodes across the entire PCB, so we have to learn how to route and connect certain critical parts of the converter in such a way that we minimize the effect of any remaining imbalances. An example of that is the connection of the voltage divider discussed below.

Note that a typical circuit schematic makes no indication or mention of even the existence of this very useful and necessary ground plane. Further, very rarely do IC designers have a precise model of the PCB built into their magnificent simulators—which may explain how they manage to prove almost any hypothesis they may be favoring on a particular day!

The Voltage Divider and Its Correct Placement

Before we go deeper into the high-frequency aspects, let us examine what happens if we ascribe even a simple DC resistance to a trace. As an example, we pick the case of a simple voltage divider, as shown in Figure 2-1. This combination of two resistors is used to set the output voltage, and is probably the most ubiquitous part of any voltage regulator—whether

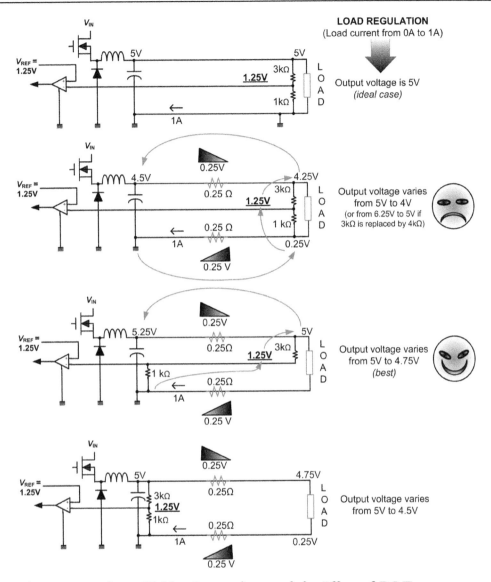

Figure 2-1 Voltage Divider Connections and the Effect of DC Trace Resistances on Output Voltage

an LDO, a high-powered AC-DC switching supply, an exotic PFC pre-regulator, or a tiny low power DC-DC tucked deep inside our cell phone. It is always there!

The question we pose here is, what happens if the load is "far away" from the converter? Note that far away need not just be in terms of an absolute physical distance. For example, that description could well apply to a typical clamshell cell phone (flip type), where the converter is located on one half and its load on the other, with very narrow traces running through intervening flex-cables and connectors. In fact, a customer had this very discussion with me quite recently. He was concerned that if the load were too far from the converter, he could lose a significant amount of regulation at the point-of-load, leading to malfunction in his cell phone. The following discussion is an explanation of all the suggestions I offered him over the phone.

In Figure 2-1 we see the various possibilities related to where exactly to connect our voltage divider resistors. We have chosen a Buck regulator, set to deliver 5V at 1A (max load) for our example.

In the first case we have the *ideal* situation, with no intervening trace resistances considered. So since the reference voltage is 1.25V, we have chosen 1kΩ and 3kΩ resistors to get an output of 5V. Note that the classical equation of a divider is

$$V_O = \frac{R_{UPPER} + R_{LOWER}}{R_{LOWER}} \times V_{REF}$$

However, more precisely:

The divider takes the voltage difference across its lower resistor (usually V_{REF}) and leverages it by the factor R_{TOTAL}/R_{LOWER} (where R_{TOTAL} is the sum of the two resistances), to give us the voltage across the entire divider (usually the output voltage).

So for the ideal case, we get as expected

$$V_O = \frac{3k + 1k}{1k} \times 1.25 = 4 \times 1.25 = 5V$$

Note that the divider's leverage factor is 4 here. Keep that in mind as we proceed to the nonideal cases below, because it is this factor that forms the common thread in all the analyses.

In the next case, we place the *entire divider at the load end* and we assume (somewhat exaggerated) 0.25Ω trace resistances as shown. Follow the gray arrows in Figure 2-1 for the discussion below:

a) Starting from the ground (which is now taken to mean the ground in the immediate vicinity of the control of the converter), we arrive at 0.25V at the lower end of the lower resistor (because of the ohmic drop in the return trace).

b) But the middle of the divider is still fixed at 1.25V, since the feedback trace carries almost no current.

c) Therefore the voltage difference across the lower resistor is 1V.

d)   With a leverage factor of 4, we get 4V across the entire divider.

e)   But since its lower end is at 0.25V, its upper end must be at $4 + 0.25 = 4.25V$.

f)   Accounting for the drop across the output trace, the voltage across the output capacitor of the converter must therefore be 4.5V.

g)   The output voltage (i.e., across the load) is therefore $4.25V - 0.25V = 4V$.

h)   These calculations were carried out at 1A (max load current). But if the load current decreases to zero, so will the ohmic drops, and the output voltage will return to 5V. In other words the load regulation (i.e., the percentage change in output voltage over the entire load current range) is $-1V/5V = -20\%$. Of course we could prefer to mentally center the nominal output voltage at 4.5V, and state the load regulation as $\pm0.5/4.5$ or $\pm11.1\%$.

We then follow the same logic to analyze the other nonideal cases, the common thread being the leverage factor of the divider.

We thus realize that *it is best to connect the lower end of the lower resistor to the converter ground, and the upper end of the upper resistor directly to the load.* That configuration gives us the best load regulation.

Any remnant error is purely attributable to the ohmic drop across the *ground trace*. So if the ground (between the converter and the load) is beefed up (e.g., a thick trace, or a nice ground plane going all the way, or an underlying metal sheet, etc.), the load regulation becomes close to the ideal case. The ohmic drop across the output trace then no longer affects the output voltage.

Efficiency Measurements and DC Resistances

Keeping DC trace/lead resistances in mind, we realize the correct way to do an efficiency measurement is as shown in Figure 2-2.

For the case labeled DC-DC Low-Power, we insert digital multimeters (DMMs) to measure the input and output current. We also connect DMMs directly across the input and output terminals of the converter board to measure input and output voltages.

In high-power applications, the internal impedance of a DMM measuring the input current may itself create problems, because its impedance may not be low enough. The contact resistance of the banana plugs leading to it could also be part of the problem. These factors may cause unintended wobble in the supply voltage at the input of the converter/control IC (this can be confirmed by a scope, if we take before and after pictures). This wobble can eventually lead to observable problems such as difficulty achieving regulation at the rated minimum input voltage, or difficulty starting up because of the high inrush current usually

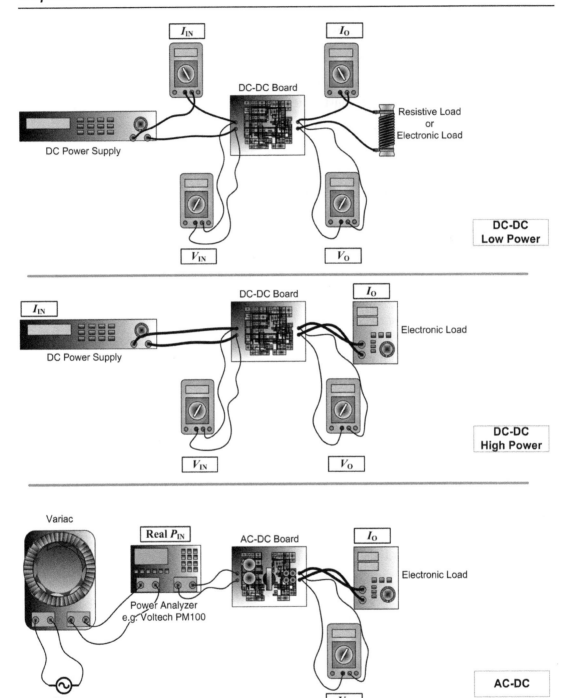

Figure 2-2 Kelvin Sensing and Efficiency Measurements

demanded by converters at power-up. In such cases, the DMM meant for measuring the input current may either have to be replaced with a much better one, or we may need to rely on the (usually somewhat less accurate) current readings of the DC power supply itself. Similarly, we may need to replace the load resistor with an electronic load, and thereby dispense with the DMM meant to read the output current (the load current can be directly read off the load display panel). Note that for most other testing, it is always recommended to use an electronic load set to CC (constant current) mode. Of course, we are never 100% sure what exact type of load profile will eventually show up at the output terminals of the converter, but it is normally assumed that CC mode testing is severe enough for most purposes. But certainly, we should never walk away saying the converter has no *startup issues* after only testing it with either a passive load resistor or an electronic load set to CR mode (constant resistance mode).

In AC-DC power supplies we typically use a variac followed by a good power analyzer (the Voltech PM100 has been an all-time favorite of engineers). The Voltech will tell us virtually everything that we ever wanted to know about the input side but were afraid to ask. For example, it will tell us $V_{RMS} \times I_{RMS}$ as V-A (the volt-amperes), or the real power displayed as *W*, or even the power factor, crest factor, and so on. Just ask! However, remember that a variac itself ends up changing the input power factor somewhat, and can also affect holdup time measurements (its magnetization energy tends to push its way through even after the input bridge rectifier of the power supply is supposed to have stopped conducting). So you may need to simply plug the power supply into a wall outlet, or get an expensive programmable AC source. The latter will also come in handy for other tests such as repetitive AC inputs with missing half cycles, dropouts, line droops, line transients, and on and on.

Kelvin Sensing

The above efficiency measurements are an example of Kelvin sensing, also called 4T or four-terminal sensing (although on a PCB, two of these terminals may combine into a single ground plane). Basically, we are trying to ensure that the current and voltage measurements don't interfere with each other because of intervening impedances. A similar recommendation applies to PCBs if we want to measure the voltage across, say, a sense resistor, for providing a more faithful current waveform for implementing current mode control, or simply for a predictable cycle-by-cycle fault current limiting. See Figure 2-3.

Source-to-Ground Trace Inductances

The drive signal applied to the switching Fet is usually created by a driver stage sitting inside the IC, and therefore connected to the IC ground. But the actual behavior displayed by a Fet is not determined by the voltage applied on its Gate terminal with respect to our

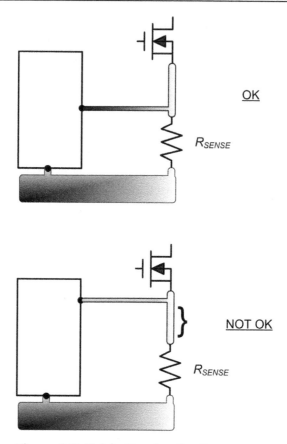

Figure 2-3 Kelvin Sensing for Better Current Sensing

ground reference (wherever that may be), but with respect to its own Source terminal. The actual V_{GS} is all that matters to it.

So if, for example, the source-to-ground trace is a little too long, it can generate a significant inductive kick at the instant of a switch transition that can at best slow down the transition somewhat, or at worst, produce spurious (unintended) turn-on and turn-off of the Fet, leading to its destruction.

In Figure 2-4 we show a rather benign case of what can happen at the instant of turn-off. The Gate commands the Fet to turn OFF, but the PCB trace impedance in the Source is also previously carrying current and tries to resist the change by creating a small voltage source (spike) that attempts to keep current flowing until its energy is dissipated. This causes V_{GS} to change shape and the transition slows down. However, this is not one of those recommended ways to slow down the transitions with, because it is based essentially on parasitics, which we know are very unpredictable.

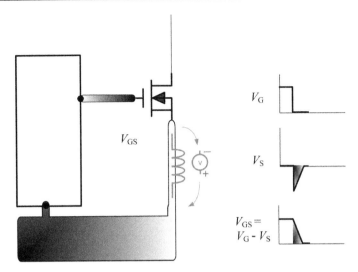

Figure 2-4 Effects of Parasitic Inductance in the Source at Turn-off

Many years ago, after I had just left a company making AC-DC power supplies and moved on to a nearby manufacturer of high-voltage power semiconductor ICs, I suddenly started hearing how my previous tech had managed to create an awesome display of bench-top fireworks, that too with the beloved 900W power supply that I myself had built and just left behind. I also heard that that series of multiple explosions had lasted at least 30 to 45 seconds, much to the joy of a small crowd of cheering engineers (standing a safe distance away). How did he manage that? I remember, up until then, the only major excitement in that lab used to occur moments after my senior-most colleague would creep up with a roll of bubble wrap just inches away from the ear of an unsuspecting engineer who was gingerly poised in the process of inserting a probe on to an awkwardly humming power supply. Finally HR (Human Resources) convinced him to bury *that* particular test. But in this recent episode, the tech had simply forgotten the golden rule of *never playing with the Source inductance of a switching Fet*. There were four paralleled Fets in that high-power PFC stage. So when the manager requested him to measure the Primary current, he quickly broke the Source trace leading to one and inserted a small loop of wire for inserting a current probe. In doing so, he also managed to break the symmetry, which was ensuring good dynamic sharing of the four Fets (I had gone through a lot of trouble laying out the PCB so that all four Fets had exactly the same amount of trace length and width in their Source paths, before connecting to the input capacitor of the PFC stage, besides fully symmetrical Gate traces too). The rest became history. First the four PFC Fets blew one by one, then all the Fets of the multiple downstream PWM Forward converter power trains erupted, then the secondary side diodes and components charred, the transformers also smoked (and predictably died doing so, just as the Doctor had warned!). Not to forget the hitherto unnoticed low-power standby supply that also decided to quit rather loudly along with the majority. Scurrilous fence-sitter! I wonder why the input fuse didn't blow! Actually it must have, but the huge amount of energy already stored in the two huge 400V bulk capacitors was enough to keep the show going for a very, very long time.

After this magnificent display, I have a feeling that their eternally tetchy prototype production department asked the morose tech to rework and fix the power supply himself ("we only make *first* articles, and you know that").

The bottom line is—to measure Fet current, *insert the current probe into its Drain*, never in its Source. Put in a small sense resistor in the Source if you must, but nothing *inductive* please.

Avoid Wirewound Resistors

Note that to avoid the inductive kick, we must try to keep the overall length of PCB trace from Source to ground as small as possible. However if there is a sense resistor present, it must be of a low-inductance type; otherwise it will itself create inductive bumps. Therefore, one big no-no for current sensing is the wirewound resistor. The only place I ever found it being used was as the inrush current limiting resistor of AC-DC power supplies, because in that location there are no high-frequency harmonics to deal with. But you will be surprised how many junior engineers initially try to use a wirewound as a high-frequency current sense resistor in their AC-DC prototypes. They learn quickly enough though!

One common use for wirewound resistors seems to be as the load for a converter. I also use that configuration when doing thermals to simulate the customer's system and for noise and ripple measurements. But rarely do I use it for anything else. I would strongly suggest you get yourself a good electronic load. But do remember to set it to CC mode (constant current mode). Because a resistor (or an electronic load set to CR mode) is just *too benign*. For example, rarely does it reveal any fundamental start-up issues.

Occupational Hazards

Let us now turn our attention briefly to an occupational hazard you may encounter while troubleshooting. The message simply is be prepared; don't let this particular hazard change the way you think or analyze a problem. Also, learn from your peers' experience; it is really not a good idea (in terms of job security) to tom-tom your skills to the top brass of the company! This story will help explain why.

A few years ago, one of my respected colleagues at the Silicon Valley design center of one of the world's biggest power supply manufacturers explained the rather hair-splitting logic of load regulation and voltage divider placement to me quite eloquently. I knew this engineer was not only quite smart, but had also acquired tremendous international experience ever since leaving his hometown in Sweden. But that day, he was in a state bordering pure shock. His newly appointed Boss (let's call him Mr. So-So) just wouldn't get it (why had he even gone to him, I wondered). Apparently, Mr. So-So kept insisting that the right approach was to place *both* resistors directly across the load—that is, the lower resistor connected at the load end too. I think Mr. So-So may have been partially confused by pictures of Kelvin sensing. But that technique is itself only *partially* applicable to our particular case. If we really want to do full-blown Kelvin sensing (to eliminate any ground drops too), we actually need a front-end differential input amplifier to process the feedback signal first, not a simple direct connection to an error amplifier with its reference tied to the IC ground, as we had. In our particular case we can only make matters worse by connecting the lower resistor at the load end. So to be fair to Mr. So-So, perhaps all this was not so completely obvious. It wasn't to me either at that time, but at least I was willing to follow the argument through with our senior engineer. The problem with Mr. So-So was that he was not even willing to engage in the briefest technical discussion about it. Worse, every time he felt confronted, he would emphatically proffer his resumé to win the argument, saying, "We did it exactly like that in Astek for years, so just do it." Of course Astek would probably have something to say to that in the form of a detailed denial/disclaimer! The senior engineer was also struggling with the dawning realization that the man he was to report to for the next several years was for all practical purposes brain dead. He said he could also visualize his next annual review disappearing down a watery vortex. And in fact, historically speaking, it did!

Now, you may ask, how did Mr. So-So ever become the Applications Manager, and that too in one of the biggest power supply manufacturers of the world! Good question. But perhaps even more surprising, I believe that as of today, Mr. So-So *still* holds a similar position—this time at a well-known MP3 player manufacturer, qualifying (mercifully, not *designing* any more) the tiny DC-DC converters inside it. Which also means it is probably safe to buy that shiny white player after all.

They say history never repeats itself, only historians do. But you can see that the historians were right all along. Because the *same* person (let's call him "Marcos") was the one who appointed Mr. So-So as Apps Manager in *both* companies. Marcos himself eventually held the *same* position in *both* companies, as VP of Engineering (and Mr. So-So's supportive boss). But note that after their first stint together in Silicon Valley, they both were eventually laid off, since the company decided to firmly close that particular location (for lack of results, if you hadn't guessed). Mr. So-So finally settled for a "mere" engineering post at this MP3 company. Eventually, a timely introduction to higher-ups from him saved Marcos, who snuggled back into the VP slot he wanted so badly. I guess one of the first things Marcos must have done as VP was to "re-focus the organizational structure for improved efficiency." That simply meant appointing Mr. So-So back into the position of Apps Manager! The debt was now fully repaid. Looking back, it is fascinating for me to realize that these two managed to pull almost the same stunt *twice* in rapid succession, barely twenty miles apart. Did I hear somebody say, "Silicon Valley is a small place?" It seems their mutual connection actually went back decades, right up to their days together in Hong Kong. At this point you could start playing devil's advocate and argue that perhaps Marcos *really* knew how genuinely talented his all-time protégé was, since they had worked together for years. The only problem with that hypothesis was that Marcos had been a *mechanical* engineer all his life, and his knowledge of electronics (let alone power electronics) could probably be best described as just *so-so*.

We know all too well how incidents like these often play out at the expense of some talented engineer, who I would think should have been far more important in the general scheme of things. But don't forget, this time there was also a paying customer somewhere out there, who either settled for lousy load regulation or shelled out ten bucks more just to beef up his output connectors and cables. Though more than likely, in the latter case, those ten bucks ended up coming out of *your* pocket when you purchased the equipment. So, perhaps you shouldn't be so unconcerned now.

Watch the Feedback Trace Closely

Returning to the divider placement, we realize that so far we have really been talking only about how to (electrically) connect the resistors, not necessarily *where* to (physically) position them. But that is also quite important. Looking at Figure 2-5, we see that our previous best can be implemented in two ways (now termed better and best)—one where the upper resistor is physically close to the converter, and one where it is physically close to the load. In both cases, the electrical connectivity has remained the same, and therefore the load regulation should, in principle, remain unchanged. But maybe not *exactly* so! The reason for that is the highlighted trace in Figure 2-5 shown going into the feedback pin. This is a very sensitive trace/node and can pick up significant amounts of stray noise (we will talk about noise in more detail shortly). That in turn could affect load regulation, cause jaw-dropping amounts of jitter, and in the worst case, even induce oscillations. However, if we think about it, the feedback trace really becomes noise sensitive only *after it leaves the junction of the two resistors*, en route to the feedback pin. Why? Because the feedback trace in that portion of the run is surrounded on both sides by a high impedance. That makes it less capable of sinking any noise picked up here. Note that the amount of noise pickup will get worse in Figure 2-5, if we were to choose say 100kΩ and 300kΩ resistors instead of 1kΩ and 3kΩ. In general, we also always want to minimize the length of this inadvertent antenna. And the way to do that is shown in the lower schematic of Figure 2-5 (labeled best).

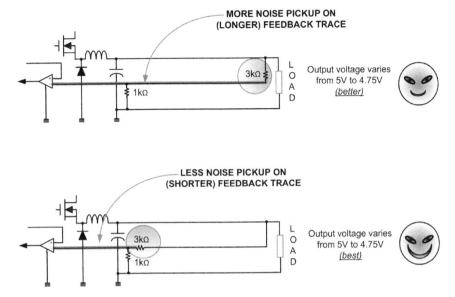

Figure 2-5 Where You Place (Not just Connect) Your Divider Is Also Important

I also like to try to ensure that the feedback trace has a *quiet* environment—perhaps ensconced between ground traces on either side, but certainly never a switching (high-frequency) trace running alongside.

Summary

We need to place both resistors of the divider physically close to the converter (and its control). The upper resistor is connected by a long trace to the output rail at the load end, whereas the lower resistor is connected directly to the converter ground (where the reference is located).

This is just an example of the complexities that can arise in debugging a simple power supply. In case you haven't guessed—this one was all about poor load regulation caused by thoughtless PCB routing!

Physical Distances Become Critical

When we come to AC effects, we have to first understand why in the example above we asked that the load be considered "far away" from the converter. We were basically trying to introduce real-world impedances to see their effect (probably somewhat exaggerated). We knew that copper being a very good conductor, we would need to be *literally* far away to create enough DC resistance for it to become significant—or to use very *thin* traces. Further, in the example above, we made it clear that those traces were only carrying DC. So, we also now realize that the above calculations would apply equally to an LDO (linear regulator), for example.

But switching regulators throw up far more serious challenges, because as mentioned previously, they have very sharp edges in their current and voltage waveforms (during switch transitions). We know that inductive impedance is by definition $L\omega$, where $\omega = 2\pi f$. We also see from Figure 2-6 that the (sharp) current waveforms of even a "lowly" 100kHz switching converter can have significant high-frequency Fourier content, running into several tens of megaHertz.

The impedance of one inch of PCB trace is typically 20nH. So its impedance to, say, a 2MHz Fourier component is

$$Z = L \times 2 \times \pi \times f = 20 \times 10^{-9} \times 2 \times \pi \times 2 \times 10^6 = 0.25\Omega$$

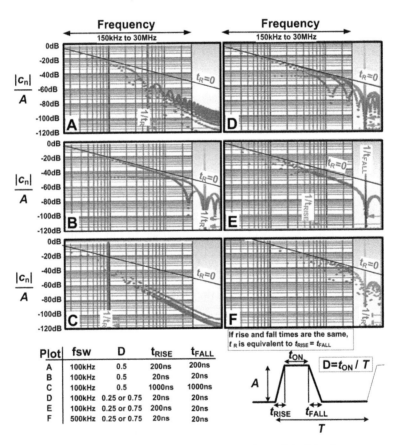

Plot	fsw	D	t_{RISE}	t_{FALL}
A	100kHz	0.5	200ns	200ns
B	100kHz	0.5	20ns	20ns
C	100kHz	0.5	1000ns	1000ns
D	100kHz	0.25 or 0.75	20ns	20ns
E	100kHz	0.25 or 0.75	200ns	20ns
F	500kHz	0.25 or 0.75	20ns	20ns

Figure 2-6 Harmonics of a Current Pedestal (with the CISPR22 Conducted EMI Compliance Region of 150kHz to 30MHz Highlighted)

We also see from Figure 2-6 that for a 100kHz square current waveform, by the time we get to 2MHz, the harmonic amplitude has typically fallen about 30dB. That means if our 100kHz current pedestal were 1A high, its 2MHz sine wave Fourier component will have an amplitude of 30mA (check: $20 \times \log(0.03) \approx -30\text{dB}$). A 2A pedestal will give us 60mA, and so on.

So for our 1A/100kHz example, the 30mA/2MHz of harmonic current, passing through 0.25Ω will give us a sine wave voltage drop of amplitude $0.03 \times 0.25 = 0.075\text{V}$—which is not completely insignificant. Therefore, our concept of what constitutes far away, starts to acquire a whole new meaning when we move from DC to AC domain. Note that in the above example, the base frequency was only 100kHz. You can imagine how important PCB distances become when the switching frequency itself is 2MHz (a number not uncommon today). For example, a 1A pedestal has a fundamental harmonic of amplitude $2/\pi$, or 0.64A. This is incidentally equal to $20 \times \log(2/\pi) = -4\text{dB}$. So in this case we get a voltage drop of $0.25 \times 0.64 = 160\text{mV}$. We realize that now even a few millimeters can be considered far away (often *too* far away). The effect on the voltage divider was just an example of the trouble that awaits us if we forget all this.

Estimating Harmonic Amplitudes

When you are on the bench, you don't want lengthy calculations interfering with your insight or thinking process. So it is always valuable to have some quick rules of thumb at your disposal (provided of course they are reliable!).

In that spirit, we now detail a quick procedure for estimating harmonic amplitudes of the square current waveform (assuming fast transitions). We will apply it to our specific 100kHz/1A example as we go along:

- Starting at the switching frequency, we need to invariably fix the harmonic amplitude at that point at −4dB. This follows from the basic governing equations (for more details, see the chapters on EMI in my "A to Z" book). Note that 0dB corresponds to the *full* amplitude of the pedestal (1A in our example).

- The envelope of the harmonic amplitudes rolls off at −20dB/decade (assuming fast transitions).

- So in going from 100kHz to 1MHz, the harmonic amplitude will fall 20dB (from its value at 100kHz). That gives us a total of $20 + 4 = 24\text{dB}$ below 1A.

- To get from, say, 1MHz to 2MHz, we keep in mind that for a 20dB/decade roll-off curve, every doubling of frequency corresponds to about 6dB of change. So at 2MHz we would have a total attenuation of $24 + 6 = 30\text{dB}$ (as eyeballed previously).

AC Resistance Is Also Important

If we have a current waveform with significant AC content, the PCB doesn't only present a parasitic *inductance* to it but an AC *resistance* too. We should be clear since these are slightly differing concepts. Basically, inductance is present only because of magnetic flux created by the current. It represents stored magnetic energy. It can therefore also give you a backlash if you try to suddenly change the current (we all know that energy cannot just be wished away; it needs to be transferred or transformed). But hypotetically, if we could somehow cancel the associated magnetic field, we would get zero inductance. This subject is discussed in more detail in Chapter 3. AC resistance, on the other hand, does *not* represent any stored energy. It is dissipative like any resistance. It is the result of the proximity effect and skin depth: as the frequency of the signal increases, it starts getting more and more confined to the surface of the conductor. Furthermore, AC resistance is also affected by stray fields from nearby current-carrying conductors. Its resultant inability to make use of the entire cross-section of the conductor results in much higher resistance at high frequencies. And that's AC resistance.

High-Frequency Input Decoupling

One of the things the preceding impedance calculations tell us is that if we are trying to draw a sharp current waveform from a capacitor, that capacitor had better be good. Because otherwise, we would be up against a very high wall of AC impedance. That will affect our ability to provide the sharp edges of the current waveform being demanded by the inductor. The current drawn from the capacitor can then have a funny rounded shape (like what we get when we do a mathematical Fourier summation limited to a small number of terms). This could induce severe *voltage* irregularities/glitches at the input pin, mainly along the edges of the switch transitions. Note that we may not even be able to see these induced voltage spikes on an oscilloscope (because they will usually be masked by stray pickup). But they are for *real*, their presence being confirmed by the *effects* they produce.

Keep in mind that the *input supply pin of any IC is one of its most sensitive pins*, because this node leads directly into almost all its control blocks. So any prevailing noise at this pin finds easy ingress into the very heart of the IC. It is hard to predict where the noise will show up finally and what its impact will be. All we can say with certainty is that the behavior *will* become uncertain! So if you see any weird behavior, just STOP. This may only be leading you down the garden path of symptomatic troubleshooting. You need to take a step back and *first try to achieve a clean supply rail for the IC*. And the way to achieve that is clearly through better bypassing/decoupling. Only if that doesn't help should you even start to look elsewhere.

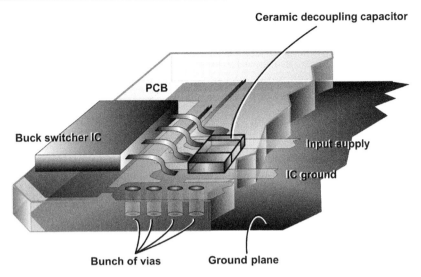

Ceramic decoupling capacitor

PCB

Buck switcher IC

Input supply

IC ground

Bunch of vias   Ground plane

Figure 2-7 A Small Ceramic Decoupling Capacitor Needs to be Mounted Very Close to the Switcher IC

Summarizing, the input (high-frequency) bypass capacitor should therefore a) have minimum inductance and internal resistance (like modern leadless ceramic capacitors), and b) also be placed very close to the switcher IC. Failure to follow these two rules can result in very erratic behavior of most switching regulator ICs. See Figure 2-7 for a recommended layout (along with the via ground stitching technique). The high-frequency bypass capacitor, as shown, is usually a 0.1µF ceramic.

Don't Forget to Place that 0.1µF Ceramic Capacitor Really Close to the IC

Until a year ago, for roughly half a decade preceding that, I used to be the final level of applications support for almost all the integrated switcher families of my previous analog semiconductor company. In particular, I was regularly fielding customer questions for their popular third-generation simple step-down switcher regulator family. These had a very fast Mosfet switch integrated on board. The transients the switch generated were known to be causing a lot of problems in the field, especially if the layout was not extremely good. But I started seeing almost half the reported customer problems magically disappear with one simple suggestion—I would tell customers to make sure they put in a 0.1µF ceramic capacitor within a millimeter of where the supply and ground pins of the IC contact the board—no intervening traces, not even an intervening via (the decoupling capacitor had to be on the same side of the PCB as the IC). This became my stock reply, before I thought about it any further (at least for this particular switcher family). Figure 2-7 is what I had been suggesting all along.

Note that I *didn't* need to insist on this 0.1µF capacitor for customers who were using the *earlier* generation of parts—mainly because those parts used integrated *bipolar* switches (BJTs). The transition time of such devices is larger—around 80 to 100ns—as compared to the 10 to 20ns for Mosfet-based switcher ICs. So the older switcher family was far more forgiving about poor PCB layout, and also generally seemed to exhibit a much lesser tendency to enter chaotic behavior (and coincidentally, less complaints from customers, too). Third-gen may really just become turd-gen for your application. Careful!

However, as you will see a little later, it is not a bad idea to *always* include this 0.1µF input decoupling capacitor. The reasons may be different on different occasions, and for different types of switchers, but this component is generally always nice to have.

You Need Bulk Capacitance, Too

Wait a minute! Apart from the problem of coping with the edges of the current transition, aren't we forgetting what will happen over the *rest of the switching cycle*? If we continue to draw say 1A for the full switching cycle from a mere 0.1µF capacitor, won't it be almost fully discharged by the time the switch turns OFF (remember it gets refreshed only during the OFF time)? We realize we need some *bulk* capacitance, too—its purpose being to keep the (relatively) low-frequency (switching frequency related) input ripple within bounds. See Figure 2-8 for a typical input noise and ripple waveform, showing both the high-frequency and low-frequency components. The high-frequency components can be tamed by the 0.1µF capacitor, and the low-frequency components need bulk capacitance.

In general, the high-frequency components are usually referred to as noise and the low-frequency component is called the ripple. Together they constitute the converter's *Noise and Ripple* (N&R), which is also sometimes called by rather weird names such as PARD (I still don't see any need to remember what *that* stands for).

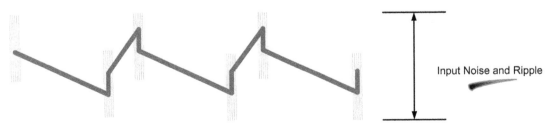

Input Noise and Ripple

Figure 2-8 Typical Input Noise and Ripple Waveform of a Buck (AC Component Only)

To calculate the value of the bulk capacitance, we need to fix the amount of input voltage ripple we can tolerate in the application. Too stringent a ripple requirement will call for an unduly large input capacitor. Of course that may be what we *want to do while troubleshooting*, but it is not a solution for mass production. So typically, for most low-voltage DC-DC converters, *the acceptable number for this low-frequency ripple component is 1% of the input applied voltage*. Of course there are several contributors to the overall noise and ripple spectrum appearing at the input pin. So let's be clear that 1% is the amount we are allocating for the low-frequency component alone. Note also that this number is still basically just a rule of thumb. A lot depends on the sensitivity of the particular IC to noise and ripple at its input. Incidentally, this threshold of chaos is rarely characterized (or at least publicized) by most IC manufacturers. For sure, it is very hard to test. So, that also means we may have to do a spot of *empirical* testing to confirm our final choice of input bulk capacitance and high-frequency bypassing.

A little later, after we have better understood the current waveforms in the input capacitors, we will do a calculation to correctly compute the amount of bulk capacitance based on the 1% input ripple criterion mentioned above.

Where Is the "Missing" Current Coming From?

Once an engineer asked me an interesting question. It was actually so *simple*, it took me by complete surprise. At that time I simply had to hem and haw my way out of the approaching quagmire, leaving some crude sketches in my trail. But I now realize that question actually leads straight into the heart of bypassing.

The engineer wanted to know what should be the minimum rating of the bench power supply for a Buck application with 5V at the input and 3.3V of output, delivering 1A of load current. I think his question initially arose because he was trying to make a small converter powered off a USB port for driving a peripheral device. He was worried how the current limit of the USB port was going to affect the load current he could draw.

I did a quick calculation and gave him a theoretical number (around 0.7A). He immediately noticed it was less than the load current of 1A, and was rather confused. "This is a Buck switcher. 1A is going *straight* into the load (from the input), so how come you are saying that the bench power supply only needs to provide 0.67A? *Where does the missing current come from?*" Simple question, but it makes you stumble, especially when you have to explain it loudly to a group of engineers leaning on every word you utter (try it!).

The math is not the problem here. That is actually very simple. But, just for fun, let's go through it once more, so we know what we are really looking for here.

The output power is

$$P_O = V_O \times I_O = 3.3 \times 1 = 3.3\text{W}$$

The input power, assuming 100% efficiency (η), is 3.3W too, since

$$P_{IN} = \frac{P_O}{\eta}$$

So the input current is

$$I_{IN} = \frac{P_{IN}}{V_{IN}} = \frac{3.3}{5} = 0.67\text{A}$$

Note that assuming 90% efficiency, we would get an input power of 3.67W. In that case, I_{IN} would be equal to 3.67/5 = 0.73A. So the actual number is somewhere around 0.7A as I told him. In reality, besides having to account for the less-than-perfect efficiency, we may need some additional margin just to provide the inrush current at startup. So it is likely we may end up using a 1A or 2A bench power supply. But that is not relevant here. The problem here is, we *still don't know where the additional current is coming from in steady state!*" Intuitively, not via math!

Actually, Figure 2-9 and Figure 2-10 put together clearly describe *everything* about the input that we need to know—how the input capacitors behave, why the load current is not the same as the input current, and also the full analysis of all the currents involved.

Let us start with Figure 2-9. Here both the capacitors have been lumped into a *single* input capacitor, one that we assume provides *perfect* decoupling. We need to follow the thick gray arrows, starting from Block 1, on to Block 2, and then on to Block 3. The logic is as follows:

- The average of the switch current waveform, *evaluated over the full cycle*, is $I_O \times D$ = 0.67A (Block 1).

- But over a full cycle, no capacitor can contribute any net DC charge (current) (for the same reason that inductors cannot contribute any net volt-seconds over a complete cycle in steady state).

- Therefore all the DC being demanded by the inductor must come solely from the bench power supply, that is, 0.67A (Block 2).

- As a corollary, the AC component being demanded by the switch/inductor waveform comes entirely from the input capacitors (only).

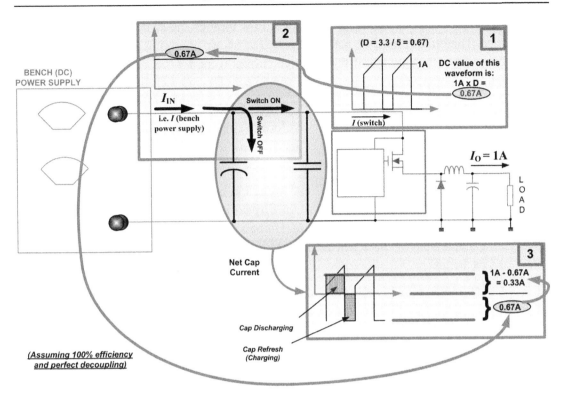

Figure 2-9 The Contribution of the Input Capacitors in a Buck

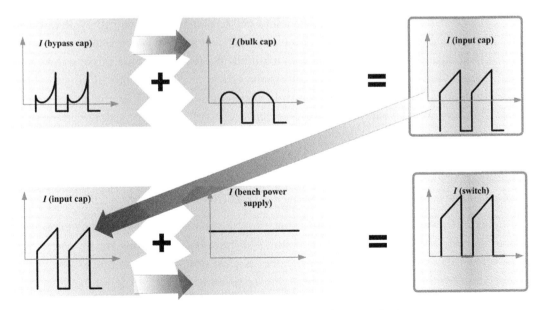

Figure 2-10 How All the Current Waveforms Add Up in a Buck Converter

- But the DC input current (0.67A) cannot flow through the switch during the OFF time. Therefore it must be getting *diverted* as shown, charging the input capacitors in the process. Note that if this diverted current were *not* 0.67A, it would amount to introducing an *AC component* in the supposedly DC input current (which in turn would imply imperfect decoupling, which is contradictory to our initial assumption).

- Therefore, if we plot the current waveform we realize that during the OFF time, the capacitor current must be sitting at a steady −0.67A (Block 3). This is the capacitor charging (refresh) current.

- However, the entire capacitor current waveform is simply equivalent to taking the switch waveform and *translating it down vertically, by an amount exactly equal to the DC value of the switch waveform.* Doing so effectively subtracts the DC component from the switch waveform and provides the required AC component to the capacitors.

- Therefore, the total height (peak to peak) of the capacitor current waveform (measured to the center of the ramp) is still 1A, which is also the case for the switch waveform from which it is derived.

- That means 1A − 0.67A = 0.33A is the *average value of the capacitor current during the switch ON time* (Block 3). In other words, although the capacitor cannot provide *net* DC (over the full cycle), it can and does provide the missing current of $I_O \times (1 - D)$ during the ON time.

- The numbers all add up finally. Both the Ampere-seconds of the input capacitor and the volt-seconds of the inductor yield a net zero value over the full cycle (a result of the capacitor-inductor duality principle). We also get our 1A current flowing to the load during the switch ON time as required. We realize that the missing 0.33A *came from the input capacitor* during the ON time. Subsequently, during the OFF time, this capacitor gets charged (refreshed), but just enough to repeat the same process every cycle. The Ampere-seconds (charge) is thereby maintained every cycle, as we would expect.

Note that in a Buck, the average input current is NOT the *load* current (as the engineer was rather intuitively visualizing), but the average *switch* current. The average switch current is of course related to the load current by $I_{SW} = I_O \times D$ (for a Buck).

In Figure 2-10, we finally break up the input capacitance into a high-frequency capacitor and a relatively low-frequency bulk capacitor. The current distributions are shown, as well as how they all add up eventually. The mystery is clear now, and in the process we also understand how the decoupling capacitors are *supposed* to behave. Now we can also start to understand how this delicate balance can be easily shattered by *lack* of proper decoupling!

Check Your Bench Power Supply

We have implicitly assumed so far that the long leads (cables) coming from the bench supply have very high impedance, so they are unable to provide *any* AC content to the converter. Alternatively, we could have assumed that the input capacitors are perfect. But neither statement is fully correct! In reality, currents will distribute according to the ratio of the impedance presented by the decoupling capacitors and the impedance of the input cables. Therefore, if we place a current probe on either of the incoming supply cables, we may get to see a lot of wobble on the supposedly DC current waveform. In effect, the bench power supply is trying to help the decoupling capacitors maintain the sharp switch waveform being demanded (a favor being returned).

But if the input supply impedances are significant, *and* the decoupling capacitors are not as good as we had imagined, the voltage at the input of our converter will start to see severe spikes, droops, and ringing as mentioned earlier. Therefore, the first thing we may need to do when a customer reports a problem is to question him or her about the exact input supply configuration. It is a hard fact that *a lot of problems seen in the lab can be traced back to some sort of interaction with the equipment providing the input power to the converter.* We should never *assume* the input rail is clean. It really never is. The question is, is it bad enough to be causing the problem? Therefore, we should *always* have a scope probe present at the input pin (close to the IC), and see what really happened there at exactly the moment the problem occurred. That may give us a vital clue. Of course this won't tell us anything if *noise* was solely responsible. For that we need to actually put in a small ceramic capacitor and see if the symptoms are alleviated.

Therefore,

- If the high-frequency decoupling is poor, the only way to check for it is to put a 0.1μF capacitor right next to the pins of the IC and see if the problem goes away. There is almost no way we can ever really *see* the cause of that sort of problem on any of our instruments. We have to *deduce* it.

- To check if our bulk decoupling is good enough, we can place a current probe on the leads coming from the bench supply and check if the wobble seems too much.

- But we must also confirm that the bench power supply is not the problem, if possible by swapping it with a completely different one (brand/rating) lying around the lab.

- We must confirm that the (low-frequency) input voltage ripple is less than 1% (i.e., ±0.5%!).

Lack of Bulk Capacitance and/or Too Much ESR Can Play Havoc

We know that a 0.1µF input capacitor takes care of the (high-frequency) noise. But it neither *can* nor *do* almost anything to smooth out the (low-frequency) ripple. However, we are now in a position to start calculating how much bulk capacitance we really need to ensure trouble-free performance (for *typical* ICs!).

Example: A Buck IC switching at 2MHz with 9V input is to provide 5V @ 1A. How much input bulk capacitance is required?

Look at Figure 2-11 where we show two cases. One is the simplified low-frequency ripple waveform (in black). The partially overlapping thick gray waveform is the more real-world situation where the input capacitor has some ESR. But let us first do the simplified calculation.

We pick the OFF time for the following calculation since the capacitor current is relatively fixed during this interval (and so we can therefore truly apply the equation $I = CdV/dt$).

The duty cycle is 5/9 = 0.56. The time period T is 1/2MHz = 0.5µs. So the bulk capacitor sees a current of $I_0 \times D = 0.56$A for $(1 - D) \times T = 0.22$µs. The amount of ripple we are allowing is $1\% \times 9$V = 0.09V. Therefore

$$C = I \times \frac{dt}{dV} = 0.56 \times \frac{0.22 \times 10^{-6}}{0.09} = 1.36\mu F$$

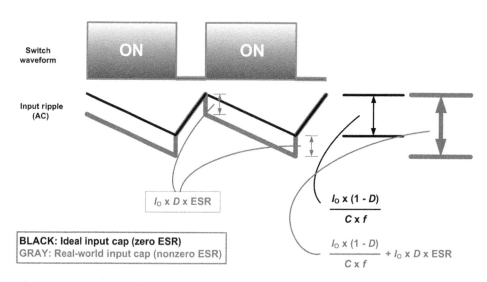

Figure 2-11 Input Ripple of a Buck Converter Showing the Effect of a Non-ideal Bulk Capacitor

In reality, since most ceramic capacitors have a real-world capacitance almost half of what their manufacturers would have you believe (more on that later), we probably need at least a branded 2.2µF or even a 3.3µF ceramic capacitor. Also note that in the above calculation we have neglected the drop across the ESR of the input capacitor. In general, that adds a little edge to the input waveform as shown in Figure 2-11 (gray waveform). So we can observe from Figure 2-11 that the calculated low-frequency ripple increases from the zero-ESR case described above, by an amount equal to $(I_O \times D) \times$ ESR (not *twice* that amount, as in fact several engineers still believe—just stare at the figure more carefully!). Note that ceramic capacitors have very low ESR (typically 5–30mΩ). So for a 30mΩ ESR, the ripple will increase by an additional $0.56 \times 0.03 = 17$mV (our overall target being less than 90mV). A little more ripple than targeted may still be OK, but if we have a typical *electrolytic* input bulk capacitor with, say, 200mΩ of ESR, we will get an ESR contribution alone of 111mV, which is already way beyond our budget.

You may need to parallel several aluminum electrolytics to lower the ESR sufficiently, *and* you may also have to substantially increase the capacitance just to stay within the total 1% limit somehow. Also remember that the ESR of aluminum electrolytic capacitors gets significantly worse over time. So *if you have a customer return after several months in the field, it may well be because of the aging of the electrolytic bulk capacitor*! Try replacing the capacitor and then recheck.

In general, *we always need to put a scope probe on the input pin of a switcher IC and confirm that the ripple is within 1% of the input voltage*. Otherwise chances are high the control sections will exhibit weird behavior. If not on one prototype, on another! Without an actual measurement, you will never know if the problem is just waiting to happen.

So if you are putting capacitors at the input without much thought (hoping it works, with no prior calculations or subsequent measurements), you can be the next easy victim of inexplicable chaos.

Question: Your calculated bulk capacitance for a certain application is 45µF. Would you pick a 47µF capacitor for the purpose?

Answer: Typically *we always need to pick a capacitor of at least 1.4 to 1.5 times the theoretically calculated value*. A lot of factors can affect the *actual* capacitance a capacitor presents in a given application at a given time. The capacitance provided by most ceramic capacitors, for example, reduces significantly if the *applied* voltage increases. Aluminum electrolytics have more stable capacitance on paper, but they show significant aging (as their electrolyte dries up). They lose a lot of capacitance eventually and their ESR also increases. (Read Chapter 4, titled "Using Capacitors Wisely"). On the other hand, tantalum capacitors have a fairly acceptable ESR, a stable capacitance, no aging phenomena, and wide operating temperature range. But they are restricted to a maximum 50V rating, which in a real application actually becomes about only *half that value*. And they erupt rather noxiously as you may know by now. So in our example, a suitable capacitor could be a ceramic 68µF. With aluminum electrolytics, because of the higher end-of-life ESR, we may need even two 47µF capacitors in parallel, or a single low-ESR type of value around 100μF. For tantalums, a single 47μF/35V capacitor is likely to suffice in an application where the input is no more than 18V.

We realize that being a pure theoretician would clearly spell trouble here. Remember that in the early days of AC-DC switch-mode power supplies, the biggest initial surprise (and mass recalls) were attributable to the aluminum capacitors losing their capacitance (or even exploding/venting in the field). But remember that for such capacitors, it is not just enough to pick a larger value to start with, but to carefully calculate (and then *measure*) the *RMS current* passing through it, and then follow the vendor's recommendations on how to maximize the capacitor's life. It is never very obvious when picking aluminum electrolytic capacitors for converters whether the final choice would be dictated by its RMS current capability or on the ripple voltage it produces (its capacitance and ESR). Both criteria, however, need to be ultimately satisfied. In AC-DC supplies, holdup time may be the clincher.

Single Ceramic Capacitor for Both Noise and Ripple?

We can ask, why not just use a single *ceramic* capacitor to satisfy *both* high-frequency and low-frequency bypassing requirements? That is certainly possible in some cases, but remember that a small-package ceramic capacitor has much lower internal impedance at higher frequencies than a bigger ceramic one. Its resonant frequency (beyond which it starts appearing like an inductance) is also much higher (a typical 0.1μF capacitor is effective *up to 30MHz*, a 1μF capacitor *only up to 10MHz*). Further, from the point of view of PCB layout, it is easier to position a small package right next to the (usually small) IC package, and with shorter traces, thus really succeeding in lowering the impedance between the input and ground. That is why the prevailing solution today is usually a small ceramic capacitor *in parallel* with a larger (bulk) capacitor. The latter may or may not be ceramic, depending on the application and switching frequency.

Note that since the impedance requirements for the bulk capacitor are relatively less stringent, it can often be placed up to a convenient centimeter or two away without causing any issues. Note that placing these input capacitors too far apart will create a C-L-C π-filter with the intervening trace inductance and that can create its own resonances and/or ringing. But done right, we will free up some valuable space in the immediate vicinity of the IC for other, more critical, component placements. Of course if we decide to move the bulk capacitor *away* from the IC, whether it is ceramic or electrolytic, or if we use a bulk capacitor of higher internal impedance even if it is close to the IC, we *must* then place a 0.1μF ceramic capacitor very close to the IC (to cater to the edges of the switch transitions).

Question: Can I improve the bypassing by several identical capacitors in parallel?

Answer: Yes, but only to some extent. The overall ESL (equivalent series inductance) and ESR are reduced, the capacitance increases proportionally, but *the self-resonant frequency of the combination capacitor is* *unchanged* from that of the single capacitor (i.e., the capacitor still starts appearing inductive above the same frequency). The governing equations are

$$C \Rightarrow nC$$

$$\text{ESR} \Rightarrow \text{ESR}/n$$

$$\text{ESL} \Rightarrow \text{ESL}/n$$

$$f_{\text{resonance}} \Rightarrow f_{\text{resonance}}$$

where n is the number of paralleled capacitors. So, if the switcher family calls for excellent high-frequency bypassing, a 0.1μF capacitor is usually unavoidable, irrespective of the bulk capacitor (or paralleled bulk capacitors) being used. A ceramic 0.1μF capacitor usually has a self-resonant frequency of 30MHz, considered adequate for most modern switchers. Too high a capacitance will have a much lower self-resonant frequency, and too low a capacitance will likely have too little energy to feed the current spikes with.

Clean Up the Supply Rail First

While commencing troubleshooting, I often solder a few identical capacitors on top of the bulk capacitor, just to improve the filtering and see if it helps (see Figure 2-12). I might also do the same for the high-frequency bypass capacitor. This technique is particularly handy when real estate is at a premium. However, to get it to work, you will probably need to apply solder fairly copiously along the sides of the capacitors. Only then does the upper capacitor really start presenting a low enough impedance at high frequencies.

Note that a quiet supply line almost always helps in debugging, whatever the problem being investigated. And luckily, enhancing the input filtering rarely causes any problems. So I

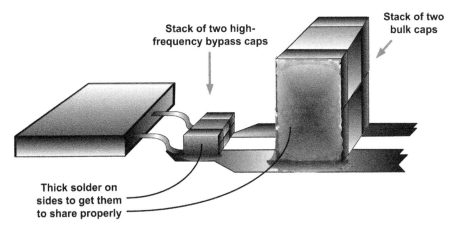

Stack of two high-frequency bypass caps

Stack of two bulk caps

Thick solder on sides to get them to share properly

Figure 2-12 Improving Input Decoupling by Stacking Capacitors

often leave these additional filtering input capacitors in place during all the debugging that follows.

In AC-DC power supplies, we usually have a controller and an external switch. Therefore we may need to worry *separately* about the bypassing requirements of the IC and the power stage. I remember I used to try and first clean up the supply rail to my UC3842 controller IC, by roughly doubling the small electrolytic capacitor present at its supply pin. The only problem with this was that the 3842 would take a rather long time to charge up the first time (through its large start-up resistor connected to the high voltage DC rail). But once it kicked in, the auxiliary winding would take over and keep the capacitor fully charged (and with less noise and ripple).

I usually prefer to remove any additional input filter components only after I have solved whatever problem I had been going after. In the case of the vertically stacked capacitors, for example, I would then start removing each additional capacitor one by one (with the help of *two* soldering irons applied simultaneously to both sides of the stack), confirming, at each step, that the problem had not come back, and that there was not a new one now!

The Control Also Needs a Clean Supply Rail

We should also remember that the pristine (sharp) current waveform being demanded by the inductor is only one of the reasons for decoupling/bypassing. The *control sections* of the IC also demand current. Though you can argue that the IC supply current as measured by a DMM (digital multimeter) is very small, in reality, it can actually have sizeable AC content too (current spikes for very short periods). During a typical switching cycle, various transistors or circuit blocks inside the IC may need to switch suddenly, comparators may change state, driver stages may demand bursts of current, and so on. In effect, all these lead to sudden changes in the IC supply current demand. Therefore, we realize that in the case of a typical Buck switcher IC (i.e., one with an integrated Mosfet and one input supply pin), the $0.1\mu F$ ceramic capacitor actually does *double duty*—it provides high-frequency decoupling/filtering to the power stages and also to the control sections. If the control and switch are physically far away, we will need to separate their filtering requirements. Note that the same high-frequency decoupling arguments apply to the Buck-Boost too, because its input current waveform is also choppy. However, the Boost is different, as we shall explore in the next section.

Boost Topology Decoupling Is Slightly Different

In a Boost topology, there are no edges of inductor current waveform at the input. That is because there is an inductor present in series with the input, which helps level any current variations. So though a certain (small) amount of bulk capacitance is still required to smooth out the slowly undulating inductor current further, *in principle*, high-frequency

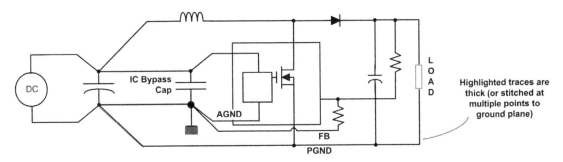

Figure 2-13 Trying to Indicate on a Schematic How a Boost Is to Be Routed

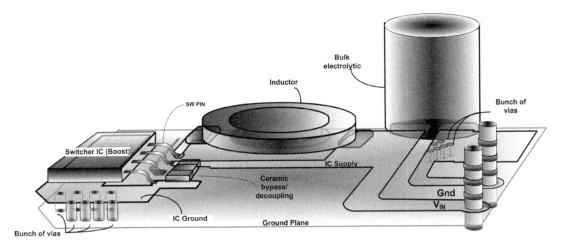

Figure 2-14 The Actual Implementation of the Input Routing Scheme of a Boost Converter

bypassing doesn't seem necessary. But wait! If we think of the control sections, we realize they certainly still need their quota of current spikes! Typically, since the control sections derive their power from the input rail, we end up putting in a small ceramic capacitor anyway, between the input rail and ground (though it could be as small as 10–22nF for a typical Boost switcher IC). Note that this capacitor *must always be positioned very close to the control sections (IC),* not, for example, at the point where the input enters the board. Its real purpose is to bypass the control stage, not the power stage. This is an important consideration when we start creating a PCB for any Boost converter. Take a look at Figure 2-13 (the schematic) and Figure 2-14 (its PCB implementation) to see what I mean.

Incidentally, you will also note that in Figure 2-13, we are finally trying to get our typical lying (cheating) schematic to tell the full story! That seems to be the only way out so far, till someone has a better idea (and hopefully reinvents the very concept of a schematic!).

Output Noise and Filtering

Measuring Output Noise and Ripple

Ultimately, the power supply is only part of a larger system. Therefore, besides being concerned about the effect of noise and ripple on the converter itself, we need to worry about its effect on the rest of the system. The good news is that if the system were excessively noise sensitive, no one would have touched switchers with a ten-foot pole (or a 10dB zero) in the first place. They would have been using those low-noise, power-guzzling LDOs (linear regulators) instead!

When a customer comes back and complains of too much output noise and ripple on his or her switching converter, pay credence to the *ripple* if you may, but the noise could simply be an artifact of his or her measurement technique (or lack thereof). Do ask the question! You may be surprised to find that a fairly large percentage have measured the ripple as in Figure 3-1. Or worse, as shown later in Figure 3-3! Notice that a large radio-receptor pickup coil has effectively been created by the scope ground lead in the first figure, and a huge radiating antenna of circulating current in the second (if a voltage differential exists along the ground plane, it will certainly drive a circulating current here). If you don't believe me, connect your probe as shown in Figure 3-2. No, this figure is not a misprint. It is slightly *different* from the preceding figure! You will be surprised to see what all you are picking up from the immediate environment. In fact, in the course of any troubleshooting, *whenever you see an odd scope picture, first try to ensure it really is real by using this simple grounded probe technique.*

The correct way to do a noise and ripple measurement is shown in Figure 3-4. Further, if noise needs to be captured, the scope should be set to AC-coupling, and the termination impedance should be set to 50Ω. Otherwise the 50Ω impedance of the coaxial cable of the probe will not match the input impedance of the scope, and signal reflections will occur along the cable. As a result, the noise amplitude apparently seen on the scope will be roughly twice what it really is. But for the ripple component, the 1MΩ termination impedance setting should be selected. Also generally speaking, for a ripple measurement you would prefer to set your probe tip on the 10:1 setting, but for noise, a 1:1 setting is more desirable (all this will be explained later).

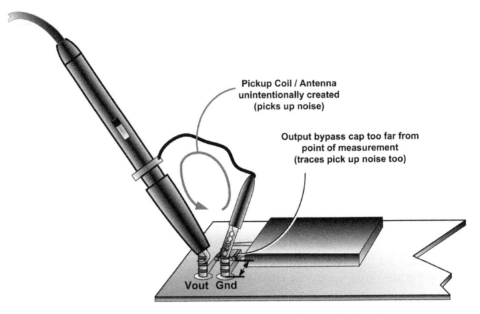

Pickup Coil / Antenna unintentionally created (picks up noise)

Output bypass cap too far from point of measurement (traces pick up noise too)

Vout   Gnd

Figure 3-1 Wrong Way to Measure Output Noise and Ripple

V_{OUT} Gnd

Figure 3-2 Checking Extraneous Noise in the Immediate Vicinity

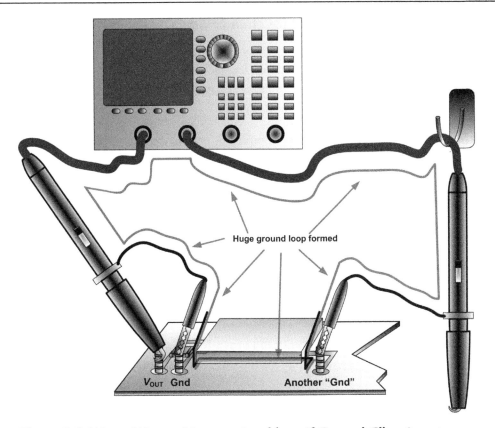

Figure 3-3 Worst Way to Measure Anything—If Ground Clips Are at Different Grounds, this Can Create a Huge Loop Antenna of Circulating Current

Figure 3-4 Correct Way to Measure Output Noise and Ripple

Output Noise and Ripple, and Its Relation to Input Noise and Ripple (PSRR)

Engineers are known to beef up the input capacitor for various reasons. At least some are hoping to reduce the *output* noise and ripple! That actually does work in some cases, but the arguments are subtle and need to be qualified, depending on the type of converter and its application. Let us go through the reasoning.

- In a DC-DC converter the ripple component travels through the power stage including its effective LC filter, to the output. So the relevant question is—what fraction of the input *ripple* reaches the output? We can look upon this process as a product of two cascaded gain stages. First the input ripple goes through the power stage. Here it either gets attenuated or enhanced, as per the ratio V_O/V_{IN} (i.e., D for a Buck, $1/(1 - D)$ for a Boost, and $D/(1 - D)$ for a Buck-Boost). The LC filter that follows attenuates this incoming signal before it reaches the output. Note that for the Buck-Boost and Boost, the L of this effective LC filter is actually $L/(1 - D)^2$, where L is the actual inductance being used. C is always the actual output capacitance. To understand this better, see the "Line to Output Transfer Function" section in my "A to Z" book. Note that at the switching frequency, the open-loop gain is very low, and so its ability to correct for the ripple component is almost nonexistent. We can virtually ignore the effect of the Bode plot on the ripple.

- Note that for both DC-DC converters and AC-DC converters, high-frequency refers to the harmonics created by the switch transitions. In a DC-DC converter, the low-frequency component (i.e., ripple) refers to all the components related to the *switching frequency*. But in an AC-DC supply, low-frequency (i.e., ripple) refers mainly to the *line frequency* component (i.e., the rectified 50 or 60Hz). So in AC-DC converters, the LC-attenuation, as explained above for DC-DC converters, applies. But we also now get significant help from the Bode plot because the Bode plot gives a very high gain at line frequencies. That helps the AC-DC converter significantly. The math is actually very simple. If the attenuation from the power stage and LC filter combined is, say, 23dB, and the DC gain in the Bode plot is 60dB, the total attenuation of the 50/60Hz ripple component is $60 + 23 = 83$dB. A switching frequency-related component would have received only 23dB of attenuation (no help from the control loop). If you think about it, this is the natural result of the standard procedure of having the open-loop gain crossover somewhere *between* DC (0Hz) and a certain fraction of the switching frequency (fsw/4, fsw/6, fsw/10, fsw/20, etc.). The converter thus ends up with high gain at DC (even at 50/60Hz), and therefore a much higher ability to correct for it at the output. But it also gives us a very small gain at the switching frequency, and therefore almost no corrective ability for the ripple component of DC-DC converters, for example.

■ The Bode plot clearly has a big impact on the output ripple in an AC-DC converter. Due to that unexpected help, we can allow significantly higher ripple at the input of its power stage—typically 15–35% of the peak-rectified voltage. Of course that helps significantly in economizing the high-voltage bulk capacitors, too. However, since such a high ripple will almost certainly throw the control circuit into jitters, we invariably need an *RC* filter at the input of the control IC.

■ The overall ability of a power supply to attenuate disturbances at its input is expressed as its PSRR (power supply rejection ratio). In graphs, PSRR is usually plotted as a function of frequency. We will invariably find that the rejection ratio is *very low at higher frequencies*. One reason for this is that the Bode plot cannot really help because the open-loop gain is very small at these frequencies. The other reason is, even a tiny stray parasitic capacitance (e.g., across the power switch and inductor) presents such a low impedance to noise frequencies (whatever their origin) that almost all the noise present at the input migrates to the output unimpeded. In other words, *the power stage attenuation (which we had earlier declared to be V_O/V_{IN}) is also nonexistent for noise (and maybe even ripple) frequencies*. The only noise attenuation comes from the *LC* filter (hopefully).

■ We therefore have several options to reduce the noise level at the output.

    a) Kill it at its *point of entry* into the food chain—that is one reason why the 0.1μF capacitor may be added on the *input pin* irrespective of the switcher family.

    b) Attenuate it *on the way* to the output—that would require an inductor with very low parallel parasitic capacitance (not very practical usually). Also there are various EMI suppression techniques as presented in Chapter 11.

    c) Kill it directly *at* the output—so we would need very good high-frequency bypassing at the output, too, though that could throw the system into *oscillations* (discussed below). We could also use *LC* post-filters if space and cost permit (described later).

    d) Reduce generation of noise at its *point of creation*—that is a good option, used in conjunction with other methods. We will discuss it further under PCB layout in Chapter 6.

Settle for Noise, not Oscillations

Yes, we could simply place a high-frequency ceramic capacitor directly across the output to kill the noise appearing there. But remember, many switchers (like the BJT-based switcher family mentioned previously) are actually relying on some *minimum* ESR at the output to make their control loop stable. Putting in a ceramic capacitor across the output is often a clear no-no!

In such cases, we could try to reduce the high-frequency output noise by suppressing it *at the input*. So that could be a valid reason to place a small ceramic capacitor at the input of an older-generation switcher IC (i.e., one with a BJT switch). Its primary purpose is then *not* to ensure that the control does not go into chaos because of switch transient noise, but to reduce the *output* noise in noise-sensitive applications.

Most of the evaluation boards of such ESR-sensitive parts are shipped out to customers with only aluminum electrolytic or tantalum capacitors at their outputs. But what *really* happens is that the customer happily connects the eval board (rather expectantly) into his or her system, and completely forgets there are a bunch of ceramic capacitors all over the system board (for local decoupling at different points). In effect, the switcher can lose that valuable zero in its control loop and break into oscillations (see Figure 3-5). More so if the connecting leads are short.

One of my major customers reached this point of no return just a few weeks before scheduled full-scale production. He was dumbfounded to see the tiny phase margins of the switcher, after I somehow convinced him it was important for him to let me hook it up to his *actual system board* and then do a proper loop check. He had not even *thought* that stability could ever become an issue. And so far, he had been very comforted doing normal functional tests using dummy loads (which are nothing more than banks of goody-goody resistors). However, now, after telling him he had a problem, I also had to help him save the show.

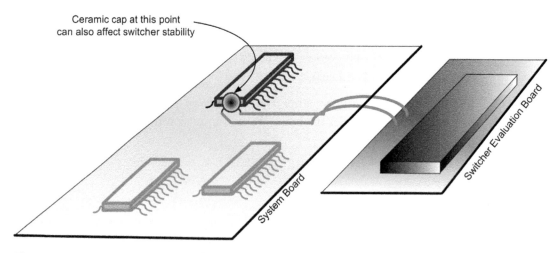

Figure 3-5 System Input Impedance Can Make the Switcher Unstable

I told him to try moving over to a switcher IC, which allowed a low-ESR capacitor on the output. But that is a major PCB redesign and complete "re-qual" (horrified looks). Alternatively, I suggested, we can try to send in a *sub* for the missing zero (a *substitute* zero or sub-zero of course!). That can be done by means of a feedforward capacitor as shown in Figure 3-6. But that solution obviously requires an external voltage divider (he had a fixed-voltage option switcher). Or we can insert a small post-*LC* filter as shown. In this case, a very small *L* usually suffices—typically around 0.1 to 0.5µH. But we have to ensure that the feedback trace still goes from the switcher side of this *LC* post-filter as shown in Figure 3-6.

When designing a PCB for a switcher IC, try to plan ahead and leave the option of moving back from a fixed-voltage part to an adjustable-voltage part (leave room for two resistors). That is a likely retreat in the face of various problems you may encounter.

Test your switcher's phase margins with something very close to the final load as early as possible. As mentioned previously, even simple startup problems of switchers do not usually show up with resistive loads. For these, you really need to test using an electronic load, placed at least in constant current (CC) mode.

The customer went with one of the suggested solutions, but I don't think he ever closed the loop completely with us on that. He was just too shaken up by his last-minute brush with death. My *assumption* is he went into mass production very soon (with minor initial rework). I have learned that in such cases, the good news always is "no news at all." And,

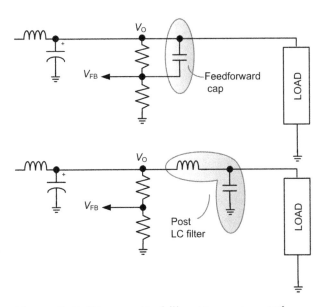

Figure 3-6 Ways to Stabilize Converters When the System Ends up Destabilizing the Switcher

as an Apps engineer, it is usually more than enough if you don't get to hear about it from your Boss. Though in rare cases, customers do get so fed up, they escalate it to the Boss. And sometimes to the Boss of the Boss! And that's exactly what happened in the following incident. But note, there is a slight twist in the ending!

Several years ago, I remember I had finally discovered the cause of the weird duty cycle related foldback behavior of the "third gen" switcher family over a rather lonely Christmas-break vigil. I was considering reporting it fairly cautiously (but candidly) to the rather justifiably irate customer. That's when my battle-honed Boss stepped in, sensing a major liability issue, and tried to extricate our Apps group completely from the situation on the excuse that we had provided "enough resources already" (in short, "get lost you small-time loser"). So one fine day, the now almost maliciously indignant customer tried to escalate it to the Boss of my Boss. He apparently called the corporate product support line demanding to speak to the Product Line Director himself. But unknown to him, the rather cocky "I can see the whole nine yards" Product Line Director (PLD) of our Power Management group was busy at his usual perch 30,000 feet in the air and had assigned his Product Engineering (PE) Manager as the acting PLD in his absence. So the customer's fulminations automatically got redirected to the PE manager. Too bad, because that guy happened to be not only just a peer of my Boss, but his long-time vacationing buddy too. Family friends really. So it was like complaining about the actions of the right hand to the left hand, with the brain not having a clue as to what had transpired below. A little after the "complaint," the PE manager strolled in nonchalantly into our lab to tell me what had happened. But he dismissed it as a "small time socket anyway," and said I shouldn't need to worry about it. However, all that changed just a few months later, when some really big customers started moving in on us in droves, reporting the exact same problem, and demanding immediate explanations. That's when we scrambled as an enviable team to put in a +400-word long disclaimer somewhere deep inside the datasheet. Clearly, a visibility of 9 yards was not enough from 300,000 feet up!

Too Much Noise? Try Slowing own the Mosfet

We know from Chapter 2 that the harmonic amplitudes depend on the rise and fall times. That is one reason why engineers often try to slow down the Mosfet (increase its transition time), usually at the expense of some efficiency, though sometimes it can even help improve the efficiency, as we will see.

This is actually a good place to talk briefly about some of the general reasons *why* we may want to slow down the power switch. And *how*.

For example, in Figure 3-7, we have a typical AC-DC power supply (Flyback or Forward). The threshold voltage of such a high-voltage Fet is typically high (around 8V). It is common to try to achieve *fast turnoff but relatively slower turn-on*. The concern is that if

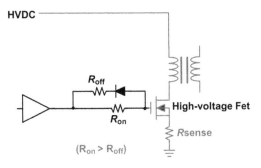

Figure 3-7 Slowing Down the Turn-on in AC-DC Applications by Increasing R_{ON}

we turn ON too fast, we may not be giving enough time for the ultra-fast output diodes to recover. So a significant diode reverse recovery current will pass though the switch, increasing its dissipation, rather than improving efficiency. Therefore typical values of R_{ON} and R_{OFF} are 22Ω and 4.7Ω, respectively. However, I remember when I was struggling to get my AC-DC power supply to meet the usual EMI standards, my far more experienced colleague redesigned the power stage. As a result, he could get by with a simple, single-stage EMI filter, with a size at least half of what I had. He didn't even have a common-mode choke anymore. Wow! Probably the most notable thing he did to achieve this was to simply increase R_{ON} to a whopping 100Ω. It worked for him! The power supply efficiency was surprisingly almost at par with mine, if not *better*. You have to remember that the EMI filter too is a major contributor to total losses. And large filters are likely to be more dissipative, too. Of course, we are assuming that the slightly increased losses in the Fet did not call for bigger heatsinks or other components.

I have described what is, in effect, the process of *leveraging efficiency*. That is, *you consciously give up a watt or two here, just to gain much more elsewhere.*

Another example of this is the standard PFC pre-regulator Boost converter shown in Figure 3-8. If the switch turns ON too fast, it invariably leads to a huge reverse recovery current through the diode and switch. Not only does that reduce efficiency by 5 to 10% typically, but since this current spike has very high-frequency content, the PCB will also complain, and we will get higher EMI, increased output noise, and ripple. There are many ways to tackle this as shown in the figure. Yes, we can always *degrade the turn-on* of the switch (increase its pull-up drive resistance), but the impact on efficiency is quite significant because of the high voltages involved (we are typically expecting 85 to 95% efficiency for such a stage in this application). So some engineers use innovative turn-on snubbers, mostly proprietary. Note that any such snubber typically works by slightly delaying the leading edge of the current waveform at turn-on, collecting the energy expended in the process in a

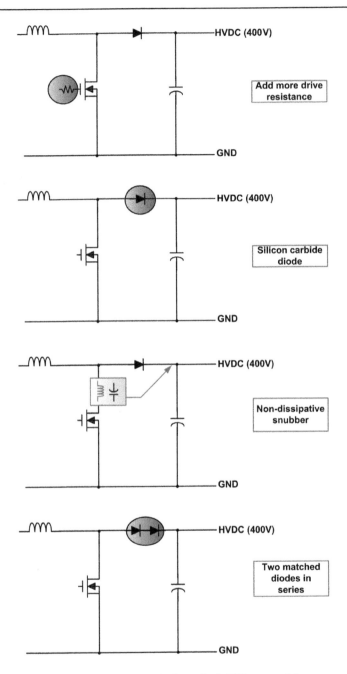

Improving efficiency and reducing noise in PFC pre-regulators

Figure 3-8 Ways to Reduce the Reverse Current Spike in High-voltage Boost Regulators

small choke, and later, when the switch turns OFF, cycling this energy into the output (or input), so it is not wasted (burned up as heat). Alternatively, some engineers are willing to pay a lot more nowadays to upgrade the PFC catch diode to a silicon carbide (SiC) type, which has virtually zero recovery time (courtesy Cree Inc.). But in the past, at least, many engineers improved the situation with the simple logic that instead of one 600V PFC diode, two in series (of roughly 400V rating each) provide much faster recovery. Usually, the additional heat lost is conduction loss because the additional diode forward drop is more than compensated for by the improvement in switching losses (of which the reverse recovery current is a major part). The only problem is to get the two series-connected diodes to share the voltage well, not only in steady state, but also dynamically (when in transition). That is not easy to ensure, so some vendors (such as ST Microelectronics) have come up with "tandem" diodes for this application, consisting of two diodes on the same chip, and therefore well-matched across production spreads. Basically, they use the same relative matching principle as the balun-drive I created for the ballast (see Chapter 1).

Another way of reducing the reverse recovery current shoot-through is simply to ensure that the boost diode is carrying *no* forward current at the moment when the switch starts to turn ON. The diode then blocks reverse voltage instantly. In other words, running the Boost in DCM or BCM (boundary conduction mode, i.e., at the critical boundary) will produce higher peak currents, but *smaller* inductors (yes, if *r* is large, the size of any inductor typically *reduces!*), and perhaps much better efficiency too, because now, the turn-on crossover loss becomes zero.

In DC-DC converters, we can somewhat slow the turn-on of Fets *if we insert a small resistor (10 to 20Ω typically) in series with the decoupling capacitor of their respective driver stages.* For example, a small resistor can be placed in series with the bootstrap capacitor of the third-generation switcher family I used to cover. That helped with almost 10 to 20% of customers, but somehow this trick didn't find its way into the applications information section of their datasheet. If the Fets are external, we can try a small resistor in series with the Gate, but this affects *both* the turn-on and turnoff (with such low threshold voltages, a diode in parallel to the resistor will not do anything).

In modern Synchronous Buck converters, one of the strong reasons for slowing down the Fets is the phenomenon of CdV/dt turn-on. If you look closely at the Gate of the lower Fet (very close to the Fet itself), you will see a small blip on it the moment the high-side Fet turns ON (Figure 3-9). In effect both high-side and low-side Fets are briefly ON simultaneously. What is happening here is that at the moment the high-side Fet turns ON, it pulls up the SW node very dramatically. This changing voltage induces a small current to flow through the Drain-to-Gate capacitance of the Fet (as per $I = CdV/dt$), and this can turn the lower Fet ON. The driver of this lower Fet (with all the intervening trace impedances) is usually too far away to respond effectively to this small blip. Eventually, this can provoke cross-conduction, which will either be totally destructive or at the bare minimum will lead

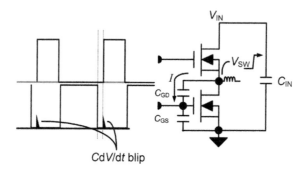

Figure 3-9 CdV/dt Cross-conduction Caused by High-side Fet Turning ON Too Rapidly

to a substantial loss in efficiency. That efficiency hit becomes especially noticeable when the converter is in normal Synchronous mode (forced PWM mode, *not* cycle skipping mode) at very light loads. My usual test is to benchmark the zero-load supply current for a good board, and then I can easily detect excessive cross-conduction if I see more than a few mA in excess of that level. I also like to compare low-side Fets during the initial selection process, in terms of the ratio C_{GD}/C_{GS} (equivalently C_{RSS}/C_{ISS}). A lower ratio makes the Fet less susceptible, and similarly, a slightly higher threshold voltage V_T improves its immunity against this spurious turn-on effect. In one IC design situation a few years ago, we actually ended up "rev'ing the silicon" (making changes to the IC itself) one last time just to make the drivers far "less aggressive." The pull-up was *reduced* by at least half, to slow down the turn-on. That also saved significant silicon area and led to a better product in general.

Once a major Taiwanese manufacturer returned his high-end laptop board to us to see why the Fets attached to our controller IC were blowing up. Of course they had put the Fets too far away. But another contributor was the fact that despite all the detail I had already put into the relevant datasheet about Fet selection criteria (the ratio C_{RSS}/C_{ISS}, for example), they had just picked whatever they wanted. Not cool! The bottom line was they had severe cross-conduction. So I asked myself, "if I can't prevent the *CdV/dt* overlap of the Gate drives, can I at least *reduce the current* that passes through the Fets during this time?" My hope was that I could somehow salvage the situation without a major PCB redesign (and potential loss of business opportunity). So in one experiment I carried out, I simply raised one end of the input bulk capacitor C_{IN} a millimeter above the board and soldered it back on with fairly thin wire (see Figure 3-10). Immediately the Fets stopped blowing up on every board I tried this on. This was almost overwhelming evidence that in the few nanoseconds of overlap, the tiny inductance I introduced was enough to restrict the cross-conduction current significantly. Of course this slight deliberate worsening of input decoupling (for the power stage, *not the controller*) *may* manifest itself as a slightly higher amount of noise on

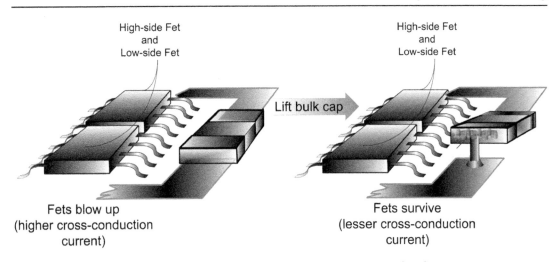

Figure 3-10 An Experiment to Limit the Amount of Cross-conduction Current

the output rails, but it still is a good compromise. Naturally, I couldn't suggest my seesaw structure as a go-to-market solution to the customer, though I did suggest a slightly higher ESR/ESL capacitor instead. However, the customer had already lost confidence in us, and the business was lost too. But I learned to advise future customers *not* to put the ceramic bulk decoupling capacitors too close to the Fets—and instead, leave slightly *longer* and *thinner* traces between the bulk capacitor and the Fets. I understand that this is contrary to most of the advice they receive from other engineers, but the problem with that advice, I feel, is that it is directed *only* at improving the output noise just a wee bit—not in the more important task of ensuring survivability under such exotic cross-conduction scenarios. The art of power supply design is the art of effective compromises, not just simple problem solving exercises. Which is also why even senior engineers sometimes end up not seeing eye to eye (preferably you-to-I) for weeks.

One thing that should have become obvious is, don't even bother to measure cross-conduction *current*. The moment you put in a small wire in series with either Fet, this current disappears. Your only chance is to put in probes to see the edges of the *voltages* of the Gate signals. But remember, you still don't know the Fet *delay times* down the road (every Fet takes a certain time to respond to an applied Gate signal, which is essentially the time it takes to charge up its input capacitance from zero to the threshold voltage through its internal parasitic gate resistor). Besides, your probes/channels may have slightly differing propagation delays. So in effect, the waveform you see may be shifted horizontally with respect to the other channel by a few nanoseconds. Only an efficiency test may truly reveal that cross-conduction may be occurring. But the current itself is probably impossible to see or measure externally.

Finally, a very important source of noise on the output (and associated EMI problems) is the high-frequency transition edge noise at the *switching node* of any converter. This is the point connecting the switch, the diode, and the inductor. The ringing can be very hard to suppress, since it involves various undocumented and stubborn parasitics—within the diode, in the inductor, in the output capacitor, and even in the switch. So if the layout is good, *and* a snubber is present across the Schottky catch diode, there is little else you can do to reduce the ringing. Yes, you can try slowing down the switching somewhat if possible, or even preferring an SMD package instead of a through-hole one for the switcher IC itself. You can also try *LC* post-filters. But eventually you may prefer a completely different IC family altogether. As mentioned previously, those with BJT switches inside are far better in terms of noise. But you usually need to watch out for that ESR zero (besides the efficiency)!

Bad Layout Can Cost You 10% Efficiency Even in Synchronous Converters

Here's how a slightly bad layout can cost you dearly—up to 10% in efficiency. In fact, as I write this, this very issue is bothering us in one of our synchronous Boost switcher ICs.

The following arguments apply to a Synchronous Buck converter, too (with the Schottky diode placed across the lower Fet), but the effects can be much more severe in a Boost because of the typically higher voltages involved.

The problem starts during the crucial deadtime marked *td* in the lower part of Figure 3-11, just when the Synchronous (high-side) Fet turns OFF and before the control Fet turns ON (i.e., the usual break-before-make protection for avoiding cross-conduction *through the Fets*). The freewheeling current has two paths to choose from, the body diode and the paralleled Schottky. Since the forward drop across the latter is much lower, *provided the parasitic inductances are low enough*, the current will prefer the Schottky during this 40ns or so typical deadtime duration. So in that case, since a Schottky recovers almost immediately, there is no shoot-through when the low-side Fet turns ON (the gray spike in the switch current waveform will *not* be there). But if the traces to the Schottky are not really *thick and short*, the current *will* prefer the body diode, since even tiny inductances can sufficiently restrict the current for this short duration. And that is when the problem starts. Once the body diode goes into conduction, it develops sufficient charge to prevent it from recovering immediately. The result is a very severe shoot-through spike through the lower Fet, just as the voltage across it starts to collapse (i.e., as it turns ON). This leads to a great increase in the turn-on crossover loss of the converter and a typical 10% fall in efficiency.

Therefore, if you know the basic reason for putting in a parallel Schottky, but don't make the layout really conducive to it, the Schottky will do almost nothing—it will be there essentially for visual and psychological appeal! Be sure to make the traces short and thick. Try removing the Schottky to see whether the efficiency improves. Because if it doesn't,

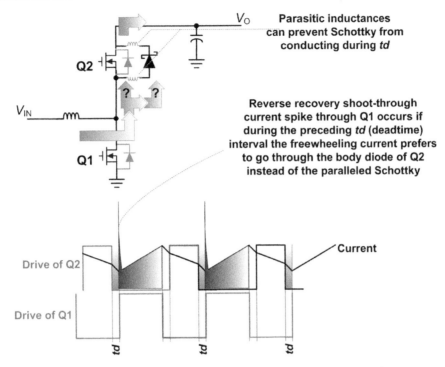

Parasitic inductances can prevent Schottky from conducting during *td*

Reverse recovery shoot-through current spike through Q1 occurs if during the preceding *td* (deadtime) interval the freewheeling current prefers to go through the body diode of Q2 instead of the paralleled Schottky

Figure 3-11 Maximizing Efficiency of a Synchronous Boost by Means of a Schottky Diode Connected with Low-inductance Traces

your layout is just not good enough. And sometimes the IC and diode bond wires will have sufficient inductance to prevent even your good PCB layout from working. In that case, the diode must be integrated in the same package (preferably on the same die) by the semiconductor manufacturer. So if you see a Mosfet with an integrated Schottky, don't think that saving you an external component was all that it is there for.

Using Capacitors Wisely

Introduction

Besides the control IC and the inductor, one of the key components used is the capacitor. It is becoming increasingly vital that we understand capacitors very well, but surprisingly, many engineers still don't. Perhaps they are taking it somewhat for granted.

A fairly comprehensive look at current capacitor technologies and their applications is provided in Figure 4-1 It is based on information easily available off the Epcos website. I vaguely remember that a few years ago, Epcos was spun off from Siemens AG (just like Infineon AG before it). Having worked for Siemens in Germany for a few years, I admit I am always interested in seeing what those guys are still up to nowadays (hopefully still sitting around guzzling beer in their *echte leder* jackets and *lederhosen*!). What I can tell you for sure is that *that* is where I met some of the most top-notch engineers ever. Of course, you sometimes had to bide your time until they sobered up, but it was well worth it! I also remember they never cut corners or jumped to conclusions (even if you *begged* them to). Quite the opposite of what seems to be the current practice in some of the bigger (and lazier) Silicon Valley analog companies.

In this particular chapter, we will focus a great deal on ceramic capacitors since these have become extremely popular today. However, in commercial AC-DC power supplies, the aluminum electrolytic (or elko) is still king, so we will talk about that component too. Unfortunately, we will have to pretend none of the others even exist. We just don't have the space for all of them here.

Part 1: Aluminum Electrolytics

Construction and Types of Elkos

Except for some exotic surface-mount technology (SMT/SMD) aluminum electrolytic capacitor types with solid electrolyte systems, in general, an aluminum electrolytic capacitor contains a wound capacitor element (the coil), impregnated with liquid electrolyte, connected to terminals, and sealed in a can (with a rubber plug at the end). The aluminum in the name,

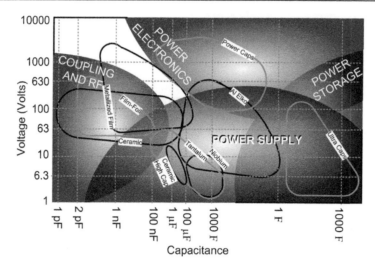

Figure 4-1 Overview of Capacitors and Their Applications

incidentally, doesn't refer to the outer can, but the *foils* inside. The element is composed of an anode foil, paper separators saturated with electrolyte, and a cathode foil. The foils made of high-purity aluminum are etched with billions of microscopic tunnels to increase the surface area in contact with the electrolyte. While it may appear that the capacitance is between the two foils, actually the capacitance is between the anode foil and the *electrolyte*. The positive plate of the capacitor is the anode foil, the dielectric is the *insulating aluminum oxide* on the anode foil, and the true negative plate is the conductive liquid electrolyte. The cathode foil merely connects to the electrolyte! This construction delivers tremendous capacitance, because etching the foils can increase the effective surface area more than a hundred times. Also, the aluminum-oxide dielectric is less than a micrometer thick. Thus the resulting capacitor has a very large plate area and a very small thickness. Remember that capacitance is given by $C = K\varepsilon_0 \times A/d$ (where K is the dielectric constant).

The development of electrolytic fluids with high levels of temperature, durability, and also rubber plugs with excellent sealing performance has led to the rapid development of SMD capacitors that can endure the reflow process (see Figure 4-2). The vertical SMD capacitor is equipped with a resin seat to ensure stability when the capacitor is mounted on the wiring board. The resin seat also protects the capacitor from heat during the reflow process (much like the natural protection given by the PCB laminate to through-hole capacitors during wave-soldering). Keep in mind that wave-soldering is sometimes called "flow soldering"; it is not "reflow soldering!" Also, don't immerse electrolytic capacitors, SMD or otherwise, into a molten bath of solder and expect them to survive. Ceramic capacitors can handle that, but not elkos.

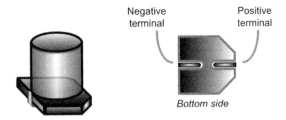

Figure 4-2 SMD Version of Aluminum Electrolytic Capacitor

Aluminum electrolytics routinely offer capacitance values from 0.1µF to 3F and voltage ratings from 5V to 500V. Note that they are *polar* devices, having distinct positive and negative terminals. They are offered in an enormous variety of styles, which include molded and can-style SMD devices (see Figure 4-2), axial- and radial-leaded can styles, snap-in terminal styles, and large-can, screw terminal styles. Representative capacitance-voltage combinations include 330µF at 100V and 6800µF at 10V for SMD devices, 100µF at 450V, 6800µF at 50V and 10,000µF at 10V for miniature can styles, 1200µF at 450V and 39,000µF at 50V for snap-in can styles, and 9000µF at 450V and 390,000µF at 50V for large-can, screw-terminal styles.

With virtually the highest available *CV* (capacitance multiplied by voltage) capability (see Figure 4-1), accompanied by the lowest cost, aluminum capacitors are still not even close to getting canned into history books, as some would think. Some younger engineers get fully charged up thinking about ceramic and modern polymer technologies, but they should also be paying close attention to the viability and finer design aspects of the still undying aluminum electrolytic capacitor, especially for low-frequency designs. Though that situation may start changing soon, due to ROHS (restriction on hazardous substances) compliance issues. Keep a watch.

A Damping Resistor (ESR) for Free

So, why not use an elko? OK, it has a higher ESR. It gives higher output ripple. Granted! It is not suitable for many low-voltage applications today. Agreed. But nothing is perfect. Surely you have seen schematics of "modern" hysteretic ICs from some companies, *with a 1Ω resistor in series with the low-ESR ceramic output capacitor*? Now, how ludicrous is that getting? At least let me ask you, don't you want the resistor *for free* sometimes? Well, it *is* present inside an elko! And the ESR helps in more ways than you can imagine. Never underestimate the role played by parasitics in power supply design. If there were no parasitics, our world would be one gigantic simulator. Either natural processes would never converge, or everything would work right off the bat (luck of the draw). So it is no surprise that all-ceramic solutions can exhibit dangerous input oscillations. It is often recommended that to damp out these oscillations we should put a high-ESR elko in parallel with the existing input ceramic capacitor. Of course we already know that a high output ESR may be required for stability too, in a converter (with either voltage-mode control or current-mode control) if its loop compensation design is not a full-blown one (i.e., Type 3 or Type 2, respectively).

Be Very Careful with Elkos

To cut to the case (oops, chase), let us therefore assume that you finally see the need to use an elko, albeit selectively. But before you plunge ahead armed with your soldering iron, I think you need to pay heed to a short story. Several years ago I had inadvertently *reversed the polarity* of a rather big, high-voltage elko on my fledgling AC-DC supply. That is a cardinal sin as I was going to find out. For a minute or so, my circuit seemed to work fine, except for an odd hum that I barely noticed (that was *normal* those days). I had no premonition that in the few seconds, I was going to be mercilessly attacked by an 85°C capacitor of very uncertain origin. Suddenly, this cheating lowlife of a capacitor erupted all over me, with regrettable assistance from my treacherous power supply itself. The cylindrical metal outer can took to the air like some newly developed Polaris missile, hitting me square on the chest and almost knocking me out cold. I returned to a state of full comprehension with a variety of oily liquids, plastic shrapnel, and soggy cellulose scattered all over my face, with a crowd of appreciative engineers cheering me on. That was *some* initiation ceremony at my new job as a Junior Research Officer at the Central R&D in Bombay, under Doc Murthy.

See Figure 4-2 for a can't miss polarity indicator of a typical SMD elko. In through-hole designs, there is invariably a long bar (or arrows) running alongside the *negative* terminal. In addition, the negative terminal lead is also typically the *shorter* of the two. Always be very careful, and double-check the polarity *after* soldering and *before* powering on.

Another thing you have to be very careful about is the fact that elkos can hold a huge amount of energy just sitting there, thanks to their enormous *CV* capability and low self-leakage. I remember in Singapore we regularly designed in a large-value 1/4W resistor (or a 1/2W just for its *voltage* rating) across the bulk capacitor of our universal input Flybacks to ensure the capacitor discharged its lethal voltage within a few seconds. There is a safety requirement for that too, though for some reason I have seen several eval boards for high-voltage switchers without this resistor. I guess they are just trying to impress you with their *low component count* (that is, if you are still alive to notice!). Because, when you power down, at some point on the falling input waveform, the converter will hit UVLO (or a starved bootstrap winding), and stop switching. Thereafter, nothing is pulling out energy from the bulk capacitor anymore, except for some leakage. So you can come back even hours later, and still receive a very nasty zap, if your hands even accidentally graze the high-voltage nodes still available. One of my elderly colleagues in Bombay had a knurled and scarred, near-paralyzed right hand. When I finally asked, he revealed he had received that years ago when he had touched a screwdriver or something across the terminals of a *bank* of high-voltage capacitors, little realizing they were still charged *days after they had been powered up*. I therefore always keep a small 100kΩ or 200kΩ resistor readily available that I can touch across the terminals of the high-voltage bulk capacitor using insulated pliers, before I do anything to it.

Elkos Tolerate Abuse

The primary cause behind most electrical failure modes of elkos is heat. That is actually a good thing, because you can even exceed its rated voltage by about 20% for a short time (about 1s), just as long as the resulting heat buildup does not catch up with you. The capacitor also usually fails to open (what did you expect if its final state is called "smithereens?"). Actually, the aluminum oxide layer has self-recovering properties, and that is why shorted failures are very rare—it can usually correct a tiny short almost immediately.

Yes, the capacitor is polar (it can stand only 1.5V reverse voltage). But if two same-value aluminum electrolytic capacitors are connected in series back-to-back with the positive terminals or the negative terminals connected together, the resulting "single capacitor" is a nonpolar capacitor with half the capacitance. The two capacitors rectify the applied voltage and act as if they had been bypassed by diodes. When voltage is applied, only the correct-polarity capacitor gets the full voltage; the other sees nothing. In nonpolar aluminum electrolytic capacitors and motor-start aluminum electrolytic capacitors, a second anode foil substitutes for the cathode foil to achieve bipolarity. Note that if you place two or more identical capacitors in series (not back-to-back) to create one effectively high-voltage capacitor, don't forget that their internal parasitic leakage resistors are not likely to be even close. So you need fairly low-value identical resistors across each of them, to force them to share the applied voltage equally.

Resonant Frequency of Elkos

We should also remember that a capacitor is only a capacitor for frequencies below its internal *self-resonant frequency*. That is the point where the impedance of its capacitive portion equals its inductive impedance. Since the two impedances always produce opposite *V-I* phase relationships (180° apart), at the resonant frequency, they effectively cancel out. The remaining impedance of the capacitor at this frequency is the ESR (resistive component). At frequencies higher than this resonance value, the capacitor behaves like an inductor (for AC components). Note that an analogous situation exists for the usual inductors we use in our converters. To guarantee an inductor's high-frequency performance, we have to ensure not only that its magnetic material is still magnetic at those high frequencies (and not too lossy), but that its parasitic (parallel) capacitance is much smaller than its inductive

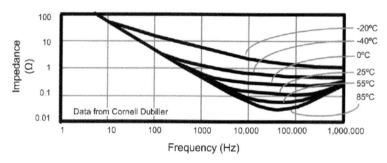

Figure 4-3 Impedance of Aluminum Electrolytic Capacitors (Frequency and Temperature)

impedance. Otherwise the inductor would also start acting like a capacitor at high enough frequencies. And it actually does, which is why noise just goes sliding past it into the output rail as discussed in Chapter 3.

The major limitation of elkos lies in the resonant frequency. A typical elko's resonant frequency is typically only 100kHz, at most 300kHz (see Figure 4-3). This means they cannot be used to provide good input or output decoupling in particular. But they certainly *can* still be used for bulk energy storage purposes (up to a certain point). They have therefore been used successfully in AC-DC and DC-DC power supplies of up to at least 150kHz switching frequency. A good thing about them is that their ESR falls both with frequency and temperature. But notice from Figure 4-3 that at high temperatures, even though the ESR (lowest point of each curve) falls at higher temperatures, the resonant frequency also falls, which if you think about it simply means that the *ESL (equivalent series inductance) increased with temperature.* So if you were counting on a high operating temperature to reduce the output ripple, for example, you might run into a surprise, since ripple is composed of both ESR-based and ESL-based components.

The biggest problem with elkos is their minimum operating temperature, because the wet electrolyte tends to freeze. Though most capacitors are rated for −40°C, and some even down to −55°C, their capacitance falls 10–20% at low temperatures (in addition to other effects). Worse, the ESR increases about ten times. So if you are counting on the "ESR zero" to provide loop stability, ensure you confirm phase margins at the lowest temperature when using elkos (actually, even the ESR of ceramics increases at low temperatures, though perhaps not that steeply).

Vibration Test Casualties

When making commercial AC-DC power supplies, you will find that large electrolytic capacitors and also transformers and inductors can easily tear off the board in any standard vibration test. That explains the generous amount of "RTV" (silicone glue) you will see all over a typical finished commercial board—its purpose to hold the heavy components firmly in place. Also, if you somehow do have the luxury of a two-sided board, learn to try and anchor all the heavy components to the board by making each of their respective PCB holes into plated through-holes (whether or not you need an electrical connection on the other side).

Life Expectancy of Elkos

As for long-term performance, the main concern with an elko is its life expectancy. Eventually, the electrolyte inside will evaporate, causing the capacitance to decrease, and beyond a certain level we would declare the capacitor "dead" (worn-out). Luckily, most vendors do not wait until it dies pathetically bleating by the wayside. They specify useful life as the point where capacitance has fallen from its *initial* value by a respectable amount of only 20% (e.g., Chemicon). But remember that the initial value can be −20% to start with (its normal ±20% initial tolerance). So the capacitance at the end of life can be 0.8 × 0.8 = 0.64, that is, 36% less than the nominal value (not 20 + 20 = 40%). So if we started with a 100µF capacitor, by the end of its useful life we would be left with a worst case of only 64µF. So if you designed your AC-DC power supply with just enough capacitance to get you out of the door, your plummeting reputation is likely to creep back in quite soon afterwards, in search of desperately needed board and lodging (*halbe pension* as the Germans call it). Things would happen! Holdup time would become inadequate rather quickly. Even the normal operating input ripple will become so high that its "average" value will fall significantly, especially at low-line. So the transistors will start dissipating more and more heat over time. The efficiency will deteriorate steadily, and also the amount of output ripple will increase, creating the possibility of system upsets. So if you have calculated that you need 100µF, you actually need to start with a value C such that $C \times 0.8 \times 0.8 = 100$, that is, $C = 100/0.64 = \ldots$. Do the math! *It is always 56% higher than your calculated value* (for 20% end of life and 20% tolerance).

It is important to double-check the failure criteria of specific vendors. For example, Panasonic SMD capacitors allow for a *30%* fall in capacitance by the end of life. That means, for a ±20% tolerance capacitor, you need to start with a nominal value *79% higher than your calculated value* (also don't forget to account for the additional fall in capacitance at low temperatures).

So what really happens if you exceed the useful life of the capacitor? In other words, suppose your power supply remains adequate in your application, by the time the 100µF capacitor hits its end-of-life low of 64µF. Can you keep counting on providence to get you past a few more years? Not really, because end of life is based on two OR-ed criteria. The capacitance can fall 20%, or the dissipation factor can increase by 200%! What is dissipation factor (DF) anyway? For elkos, it is defined as the ratio of the ESR to the

reactance (at 120Hz). So *a 100µF capacitor with an ESR of 1Ω will have a DF of ESR ×
2πfC = 0.075 at 120Hz.* Try to remember this value, and also the fact that DF is
proportional to both ESR and *C* (and frequency). That helps you to quickly figure out the
relationship between ESR and DF. Note that DF is also called the loss tangent or loss angle,
or just tanδ. Generally speaking, for capacitors, if the nominal value is equal to or greater
than 10µF, the standard measurement frequency is 120Hz; otherwise it is 1kHz. But for
elkos, this frequency is somehow always pegged at 120Hz. You may need to double-check
this in the datasheet of your specific elko. The best way of course is to treat DF as a
function of frequency, which it really is. Unfortunately, that only leads to a typical
performance curve and is not amenable to being declared as a guaranteed Min/Max in the
electrical tables of the part.

Returning to the elko, by the end of life, if we assume that *only* the capacitance fell by
20%, that would have decreased the DF by the same factor. So to get the DF to rise by 20%
from its *initial value* actually calls for an increase of ESR by the factor 200/0.8 = 250%. In
other words, by the end of its life, though the capacitance may have fallen by only a modest
amount, *the ESR has increased by 2.5 times.* So if the capacitor is being used to smooth out
high-frequency AC ripple (such as the input capacitor of a Buck and Buck-Boost, or the
output capacitor of a Boost and Buck-Boost), the heating would increase by a factor of 2.5
× 2.5 = 6.25. And what does this do? Its effect on efficiency is of course obvious, but do
not forget that this heat is the main reason for the capacitor drying up in the first place. So a
thermal runaway situation can be right around the corner—more heat, more ESR, more
heat, more ESR, and so on.

Let us sum up a few factors that play key roles in the aging process:

- The hermeticity of the end seals of the capacitor. No joint is one hundred percent
  perfect, and some evaporation will take place slowly over time. We see the need
  to pick a vendor with a high (and consistent) quality. Yes, in principle, we could
  try to seal the capacitor totally, say by immersing it in a bath of epoxy-resin/
  superglue for example! But such capacitors are designed to vent under high
  pressure (much like a pressure cooker). However, I must tell you that despite all
  that safety chatter, I have seen some capacitors explode furiously. Remember, it
  once happened to me in Bombay!

- The surrounding temperature. The heat could come from nearby components or
  through internal heat dissipation. If we lower the temperature, the evaporation rate
  will decrease and extend the life. We will see a little later how this leads to the
  published temperature multipliers.

- The core temperature. We expect that there will be hot spots inside the capacitor
  since we have less-than-perfect thermal conductivity inside it. As a worst case, *that*

is the temperature to consider when calculating life. In fact, the entire life expectancy calculation is reduced to accurately *predicting* the core temperature (since we cannot measure it).

- The ESR. This would certainly affect the internal heat dissipation, possibly raising the temperature and aiding the evaporation process.

- The frequency. Since ESR can be a function of frequency, the frequency will indirectly affect the life of the capacitor. We will see that this leads to the published "frequency multipliers."

- The most important datasheet parameter is the ripple current rating. This is typically stated in Amperes RMS at 120Hz and 105°C. The ripple current rating essentially means that if the ambient temperature is at the maximum rated of 105°C, we can pass a (low-frequency) current waveform with the stated RMS, and in doing so we will get the stated life. The declared life figure is typically 2000 hours to 10,000 hours under these conditions. Yes, there are lower grade 85°C capacitors available, but they are rarely used, as they can hardly meet typical life requirements at high ambients. There are also 125°C capacitors available, but with typically lower life figures. Take your pick.

Let us now try to understand what a frequency multiplier tells us. The ESR of an elko is usually stated at 120Hz. The vendor may have directly provided a ripple current rating at 100kHz in addition to the 120Hz number. If not, he would certainly have provided frequency multipliers. A typical frequency multiplier is 1.43 at 100kHz. That means that if the rating allows for 1A ripple current at 120Hz, then at 100kHz we are allowed 1.43A. This, by design, will produce the same heating (core temperature rise over ambient) as 1A causes at 120Hz. This is also equivalent to saying that the ESR at 100kHz is related to the ESR at 120Hz by the following equation

$$\left(\frac{I_{100\text{kHz}}}{I_{120\text{Hz}}}\right)^2 = \frac{\text{ESR}_{120\text{Hz}}}{\text{ESR}_{100\text{kHz}}} = (1.43)^2 = 2.045$$

Thus the high-frequency ESR is about half the low-frequency ESR. Frequency multipliers should always be used, or we will overestimate the heating and underestimate the life, possibly forcing us to move to a larger capacitor size (overdesign).

Temperature multipliers? These we have to be more careful about. And we have to clearly understand what they really imply.

The datasheet usually provides certain temperature multipliers for the allowable ripple current. For example, for the old but still well-known LXF series from Chemicon, the numbers provided are

1. At 65°C the temperature multiplier is 2.23.

2. At 85°C the temperature multiplier is 1.73.

3. At 10°C the temperature multiplier is 1.

This means that if, for example, the rated ripple current is 1A (at a maximum rated ambient of 105°C), then we can pass 1.73A at an ambient of 85°C and 2.23A at an ambient of 65°C. But *in doing so, the core temperature will remain the same* (not necessarily the life, though, as we shall soon see).

So what are the temperature multipliers really telling us? All they really tell us is how the vendor has designed his capacitor from a *thermal* point of view, or what exactly is the capacitor's core temperature. As we will see, if we stick to the RMS current rating of the capacitor (without applying temperature multipliers), we don't really need to know the details of the core temperature either. Temperature multipliers were therefore just objects of abuse by some designers in the past. This is perhaps why nowadays most elko datasheets are no longer even carrying that information.

But let us follow through with the exercise for now, as it does greatly increase our understanding of this vital component. We know that the amount of heating and the core temperature rise are proportional to I_{RMS}^2. So let us assume that in every case, the final core temperature is the same, that is, T_{CORE}. Then comparing the 105°C ambient case with that at 85°C, we get

$$\frac{T_{\text{CORE}} - 105}{T_{\text{CORE}} - 85} = \frac{I_{105}^2}{I_{85}^2} = \frac{1}{1.73^2} = \frac{1}{3}$$

We can thus solve for T_{CORE}

$$T_{\text{CORE}} = 115°C$$

This says that if we pass 1.73A at 85°C, or 1A at 105°C, the core temperature will be 115°C in either case. In fact, for most 105°C rated capacitors, we will have roughly a 5°C differential from ambient to the outer can and then another 5°C from the can to its innards (i.e., the core), giving us a total of 10°C from ambient to core.

Let us check our reasoning by confirming the 65°C multiplier

$$\frac{115 - 105}{115 - 65} = \frac{10}{50} = \frac{I_{105}^2}{I_{65}^2}$$

So the multiplier must be $5^{0.5} = 2.236$, which agrees with the published datasheet value. Therefore we see that from the vendor's published ripple current temperature multipliers, we can easily deduce his designed-in maximum core temperature.

The problem with this is that if the core temperature is at its maximum rated 115°C, the life would always just be the declared 2000 hours or so. That is hardly enough to get us through even one quarter of a year. We usually need at least about 44,000 hours (5 years) of life expectancy from all elkos used in a typical commercial power supply. So how do we get there? We do that by reducing the core temperature, thereby slowing down the evaporation rate of the electrolyte. Does this imply we should not be using temperature multipliers to increase the current? Yes, in fact it does.

There is actually another complication. It has been determined that not only is the absolute value of the core temperature important, but the differential from can to core is critical too. So if we increase the differential beyond the designed-in 5°C, the life can deteriorate severely, even if the can itself is held at a much lower temperature. But the designed-in differential of 5°C occurs ONLY when we pass the maximum specified ripple current (no temperature multipliers applied), irrespective of the ambient. Which means that as a matter of fact we cannot use any temperature multipliers at all. So, if the capacitor is rated to pass 1A at 105°C, then even at an ambient of, say, 65°C, we are allowed to pass only 1A, not 2.23A.

When the differential is decidedly kept equal to or less than the designed-in value, the life of the elko is then determined by the familiar doubling rule—every 10°C fall in core temperature (from its maximum rated) the life doubles. That is how we can finally get the required 44k hours. For example if the core is correctly estimated to be at 65°C, then the calculated life of a 2000 hour capacitor is actually $2000 \times 2 \times 2 \times 2 \times 2 \times 2 = 64$k hours.

We see that we cannot have our cake and eat it too. We can increase the ripple current (but degrading its life) by applying the temperature multipliers. Or we can increase the life (but not the ripple current) by not applying these multipliers. We just can't have it both ways!

Let us now do some life estimations, as this is the most critical issue facing the use of Elkos.

Question: If we pass the rated ripple current through a 2000 hour capacitor (no temperature multipliers applied) at an ambient of 55°C, what is the expected life (first pass estimate)?

Answer: At the rated current we can expect that the core is at 55°C + $\Delta T_{\text{CORE\_AMB}}$. Since we are passing only the rated ripple current, $\Delta T_{\text{CORE\_AMB}}$ is the manufacturer's designed-in core-to-ambient differential. So the temperature advantage we have thus gained (measured from the maximum rated temperature) is (105°C + $\Delta T_{\text{CORE\_AMB}}$) minus (55°C + $\Delta T_{\text{CORE\_AMB}}$), or 50°C. Since this capacitor provides 2000 hours at the maximum temperature, at the reduced ambient we may get a life of

$$2000 \times 2 \times 2 \times 2 \times 2 \times 2 = 64,000 \text{ hours}$$

Note that *in the above analysis* ΔT_{CORE_AMB} *canceled out* as we indicated previously. So this amounts to writing the following simple equation for life

$$L = L_O \times 2^{\frac{T_{CORE\_RATED} - T_{CORE\_APPLICATION}}{10}} = L_O \times 2^{\frac{T_{RATED} - T_{AMB}}{10}}$$

In our example T_{CORE_RATED} is 115°C, and T_{RATED} is the maximum rated ambient of 105°C. T_{AMB} is the actual ambient in our application. $T_{CORE\_APPLICATION}$ is the temperature of the core in our application, and in our example it is 65°C. Note, however, that we would have gotten the same life estimate had the manufacturer used *any other* ΔT_{CORE_AMB}. As we just saw, that gets canceled out. And that happened because we followed the manufacturer's recommendations and passed *only the maximum rated current*. If we didn't, the life expectancy equation we used above does not apply, as we will soon see.

But we still need to know the actual local ambient in the immediate vicinity of the capacitor. Nearby components may also be heating the capacitor. Therefore a common and perhaps conservative industry practice is to *cut the outer sleeving of the capacitor and to insert a thermocouple under the sleeve in contact with the metal case*. That way small air draughts don't affect the results. We then take this measured case temperature as the effective *ambient* for the cap, *unless of course we know better*. Suppose the case temperature is measured to be 70°C in this way, then the conservative estimate of capacitor life is

$$L = L_O \times 2^{\frac{T_{RATED} - T_{APPLICATION}}{10}} = 2000 \times 2^{\frac{105 - 70}{10}} = 22{,}600 \text{ hours}$$

However, we should be very clear what the *source* of this heating is. If it is heat from nearby components, the ΔT_{CASE_CORE} may not be that high and our estimate would be overly conservative.

Therefore a case temperature measurement may not suffice. We should also measure the ripple current passing through the capacitor.

The relevant points are summarized below

- Capacitor manufacturers recommend that in general we don't pass any more current than the maximum rated ripple current. This ripple current is the one specified at the worst case ambient (e.g., 105°C). Even at lower temperatures we should not exceed this current rating. No temperature multipliers should be used. Because only then is the case to core temperature differential within the design specifications of the part. And only then are we allowed to apply the simple 10°C doubling rule for life.

- If the measured ripple current is confirmed to be within the rating, we can then take the case temperature measurement as the basis for applying the normal 10°C doubling rule, even if the heat is coming from adjacent sources. Again, that is only because the case to core temperature differential is actually within the capacitor's design expectations.

- However, in direct customer communications, Chemicon has, at least in the past, allowed a higher ripple current than the rating. But the life calculation method given is then slightly different. This amounts to a special doubling rule *every 5°C*, which we will describe below using a practical example.

Question: We are using a 2200µF/10V capacitor from Chemicon. Its catalog specifications are 8000 hours at maximum rated 1.69A, stated at 105°C and 100kHz. The measured case temperature in our application is 84°C and the measured ripple current is 2.2A. What is the expected life?

Answer: Since we are passing more than the rated ripple current, we need to replace the usual doubling every 10°C formula with the more detailed formula made available to us by Chemicon. The calculation proceeds as follows:

$$L = L_O \times 2^{\frac{105-84}{10}} \times 2^{\frac{5-\Delta T}{5}} \text{ hours}$$

where

$$\Delta T = 5 \times \left(\frac{2.2}{1.69}\right)^2 = 8.473°C$$

So

$$L = L_O \times 2^{\frac{105-84}{10}} \times 2^{-0.695} = 21,000 \text{ hours}$$

Let us understand the terms involved here. The ΔT calculation above essentially says that

$$\frac{\Delta T}{\Delta T_{CASE\_CORE}} = \left(\frac{I_{APPLICATION}}{I_{RATED}}\right)^2$$

We know from the vendor that this family of capacitors was designed for a 5°C differential between case and core, and that differential is caused by passing the rated 1.69A through it. So this ΔT calculation gives us the temperature differential when we pass 2.2A through it. We then have a rise of 8.473°C rather than 5°C.

The term $(5 - \Delta T)$ in the exponent of the life calculation gives us the temperature *in excess of the designed 5°C*. Let us call this ΔT_{EXCESS}. So the life equation is

$$L = L_O \times 2^{\frac{T_{\text{RATED}} - T_{\text{CASE}}}{10}} \times 2^{\frac{-\Delta T_{\text{EXCESS}}}{5}} \text{ hours}$$

The first term with the positive exponent causes the life to increase above L_O, while the second term exerts the opposite effect. We can also see that a temperature differential from case to core *in excess of the designed value* is considered more harmful than a normal temperature differential (i.e., one that is caused by staying within the current rating). Chemicon models this excess temperature rise rather conservatively as *causing a halving of life every 5°C increase, rather than the usual 10°C*.

The ΔT_{EXCESS} term in the previous equation should be omitted if the ripple current is equal to or less than its rated value. In other words, ΔT_{EXCESS} is *not* allowed to be negative. In that case we revert to the usual 10°C doubling rule (i.e., just omit the $2^{a/5}$ term in the equation above).

Also note that capacitor manufacturers typically do not guarantee life under forced air cooling. The designer should either measure the capacitor without forced air cooling if possible, or add some judicious safety margin.

Rather than take the case temperature as the local ambient temperature of the capacitor, which is more of a worst-case calculation, we could try to actually measure the local ambient. Assume that the general ambient is $T_{\text{AMB\_EXT}}$. The local ambient near the capacitor is T_{AMB}. The procedure to factor out the heat from nearby components (i.e., heat which is not due to ripple current) is as follows:

1. Take the capacitor from the circuit board, putting it on the underside, but still connected to the circuit. In this position we can measure the temperature on its case, $T_{\text{CASE\_1}}$. This is

$$T_{\text{CASE\_1}} = T_{\text{AMB\_EXT}} + T_{\text{SELF-HEATING}}$$

2. At the same time we place an exactly similar capacitor at the position where the original capacitor was, but this has one lead missing, so it is in effect not connected to the circuit. We measure its case temperature $T_{\text{CASE\_2}}$. This is

$$T_{\text{CASE\_2}} = T_{\text{AMB\_EXT}} + T_{\text{EXT-HEATING}} \equiv T_{\text{AMB}}$$

3. Therefore, having measured the ambient in the surrounding air which is $T_{\text{AMB\_EXT}}$, we know all the required components of the temperature buildup.

4. Also note that the following equation is recommended for a more careful analysis of the ratio that exists between $\Delta T_{\text{CORE\_CASE}}$ and $\Delta T_{\text{CASE\_AMB}}$ (which was earlier stated to be ≈ 1)

$$\Delta T_{\text{CORE\_CASE}} \Big/ \Delta T_{\text{CASE\_AMB}} = 0.0231 \times \text{CaseDia}_{\text{MM}} + 0.845$$

I derived this curve-fit equation from datasheets and data provided by Chemicon. For the cases included at least, I could confirm that the above equation was accurate to within 6% for all capacitor outer diameters in the range of 10mm to 76mm. And for diameters greater than 40mm, the error from the use of this formula was less than 1%.

Part 2: Ceramic Capacitors

Construction of MLCCs

We will focus on modern MLCCs (multilayer ceramic capacitors). The basic principle behind these is very simple. Let us start with a simple capacitor with two plates. Its capacitance is proportional to A/d, and it occupies a volume of $A \times d$. Suppose we then split this available volume $A \times d$ into two capacitors, each of area A, but thickness $d/2$. Then each capacitor will have a capacitance of $2A/d$, and the overall volume will be unchanged. Now, if we parallel these two capacitors (by fine internal electrical connections), we will get a resultant capacitance of $4A/d$ *in the same volume*. By dividing the thickness progressively into finer and finer layers, we can thus keep increasing the capacitance. See Figure 4-4. But

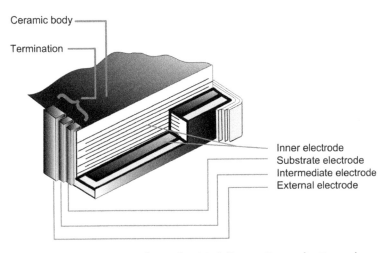

Figure 4-4 Cross-section of a Multilayer Ceramic Capacitor

at some point, each constituent capacitor will become so thin that the plates will tend to arc over through the dielectric because of the high electric field. So there is always a limit dictated by the manufacturing process and desired voltage rating.

However, the actual implementation of this basic principle takes on a staggering amount of variations in modern MLCC technology. Just like ferrite technology before it, this too has become a blend of art and science. Materials used for the dielectric vary widely, as do the internal structures.

Ceramic Classifications

Power supply designers are usually aware that the most stable ceramic capacitance comes from materials dubbed COG material, also called NP0 (for negative positive zero, referring to its near perfect temperature coefficient). But this is a low dielectric constant material, and unsuitable for modern miniaturization. So the common materials in use today are called X7R, X5R, and so on. There are others, starting with a Y or Z prefix, which no power supply designer worth his or her salt will ever use.

Anyway, let us see what this classification means:

Class I. A temperature compensating capacitor. Very stable, but will usually have a low dielectric constant (K) and therefore larger size. The most popular formulation is COG, also called NP0. The tanδ (and ESR) of COG is also relatively stable, changing by only about 25% over its full *rated* operating temperature range. COG capacitors are available from −55°C to 200°C. The ESR does increase somewhat with frequency, though the capacitance does not change significantly with frequency. COG has no aging characteristics. Note that it is actually C0G, not COG, and is an Electronic Industries Association (EIA) code.

Class II. Medium K types, for example, X7R with tanδ = 0.03, or Y5V with tanδ = 0.025 at room temperature. In both cases tanδ decreases significantly with temperature. For example, from 25°C to −40°C, the tanδ of X7R will increase by about 300%.

Class III. Even smaller than Class II (higher K) but will usually have a lower Q (higher tanδ, typically about 0.05 to 0.08 at room temperature). So the tanδ of Z5U is worse than for X7R or Y5V).

Class II and Class III are further sub-classified according to Table 4-1.

Standard Capacitor/Resistor Sizes

In Table 4-2, we have the standard SMD component sizes. Note that usually, most pick and place machines cannot mount anything bigger than size 1515. So larger components may need to be hand-soldered. For ceramic capacitors, reliability requirements call for a certain

Table 4-1 Classification of Ceramic Capacitors

Low temperature limit of range (°C)	Upper temperature limit of range (°C)	Maximum allowable change in capacitance from 25°C (at 0 VDC, over entire operating temperature range)
X = −55 Y = −30 Z = −10	4 = 65 5 = 85 6 = 105 7 = 125 8 = 150	F = ±7.5% P = ±10% R = ±15% S = ±22% T = +22, −33% U = +22, −56% V = +22, −82%

Note: A variation of ±15% (case R above) corresponds to ±150,000ppm. If the temperature range is, say, X7 (i.e., 125 + 55 = 180°C), a ±15% variation is equivalent to a TCC (*thermal coefficient of capacitance*) of ±150,000/180 = ±833ppm/°C. For COG ceramic capacitors the TCC is expressed as 0 ± 30ppm/°C applicable over the X7 range. Note that the temperature *range* is important in expressing TCC, as TCC is basically an average value.

Table 4-2 Standard Sizes of SMD Components

	Length (mm)	Width (mm)
0402	1.00	0.5
0603	1.6	0.8
0805	2.00	1.25
1206	3.20	1.60
1210	3.20	2.50
1808	4.50	2.00
1812	4.50	3.20
1825	4.50	6.40
2010	5.00	2.50
2220	5.70	6.40
2225	5.70	6.40
2318	5.80	4.60
2412	6.0	3.20
2512	6.40	3.20
2917	7.30	4.30

Note: Sizes in mm are approximate. Exact values should be calculated by converting from mils. For example, 2225 is actually 220mils × 250mils, and 0402 is 40mils × 20mils.

preheating phase even before hand-soldering. Also, because of differing thermal coefficients of expansion of the material of the capacitor and the FR4 material of the PCB on which it is mounted, it is generally not recommended to use any SMD component larger than size 2225. Since vendors have widely differing manufacturing techniques and materials, you need to confirm all these aspects from the specific vendor.

Worst-Case Variations of Capacitance

Note that Table 4-1 applies at 0V DC. In general, vendors state the nominal capacitance in a datasheet as measured at an applied voltage of $1V_{RMS}$, at 1kHz and 25°C. But in an actual circuit we could see the following *typical spreads* (after 100k hours, and also considering the possibility of both AC and DC voltages):

- *For COG*. Initial tolerance (±5%), TCC (±0.15%), voltage stability (0%), frequency stability (0%), aging (0%). Combining, we get a worst-case upper limit of $C \times 1.05 \times 1.0015 = C \times 1.0516$, that is, 5.16% higher. Similarly, the worst-case lower limit is calculated as $C \times 0.95 \times 0.9985 = C \times 0.94858$, that is, $1 - 0.94858 = 0.0514$, or 5.14% lower. The calculation is not so obvious, but truly reflects the worst-case *combinational* drift. Finally, $C \rightarrow$ *+5.16, −5.14%*. Note that we can look at the zero TCC of COG capacitors as a small nonzero TCC already included within the initial tolerance range.

- *For X7R*. Initial tolerance (±10%), TCC (+2, −10%), voltage stability (+15, −10%), frequency stability (+5, −15%), aging (−3%). Combining, we get worst case $C \rightarrow$ *+35%, −40%*.

- *For Z5U*. Initial tolerance (±20%), TCC (+2, −54%), voltage stability (+22, −56%), frequency stability (+5, −15%), aging (−25%). Combining, we get $C \rightarrow$ *+57%, −90%*.

You can see why anything worse than X7R or X5R should be avoided like the plague in power supply design. Also note that the above base percentages are still only typical (though combined). Things can get worse as we will see below.

ESR of Different Materials

Murata has a nice design tool linked to their database, downloadable from their website. However, you may have to struggle to get reasonable clarity from the tiny curves that pop up. So, I had to trace them out and redraw them rather painfully (on my noisy 10-inch Vaio, of course).

Take a look at some data I extracted from the Murata database in Figure 4-5. All these are 1μF/25V capacitors in the 1206 size, but I am varying their *material*. You see that though their resonant frequency does not change with material, the better materials have better ESRs, too. So X7R could give you an ESR almost 10 times lower. However, the difference in ESR due to material was not so obvious for lower values of capacitance (probably because then a major part of the ESR is located outside of the actual material (in the interconnects, terminations, etc.).

In any case, from now on, we will focus our attention on X7R only.

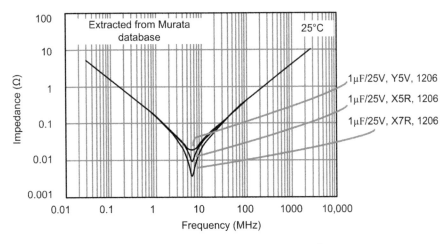

Figure 4-5 Impedance of 1µF/25V Murata Capacitors Versus Frequency for Different Materials in the 1206 Size

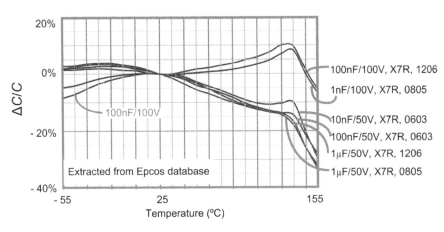

Figure 4-6 Temperature Coefficient of X7R Capacitors from Epcos

Will the Real X7R Stand up Please?

Most people assume all X7Rs are the same. They actually think it is a specific material, and that all vendors with an X7R capacitor on hand are equivalent competitors. That is simply not true. Even a given vendor can have several X7R formulations with dielectric constants ranging from 1000 to 7000. X7R only refers to a material with a TCC of ±15% over −55°C to 125°C. And that too, only with zero applied volts (or close to it). Take a look at Figure 4-6. These are curves extracted and superimposed (rather painfully) from the Epcos database of MLCCs. You can clearly see that all these are labeled X7R, but their temperature profile visibly falls into two main categories. So, if somebody says to you "the capacitance of X7R falls at high temperatures," look at him or her as if he or she just said, "all dogs are black."

Voltage Coefficient of X7R

Yes, Figure 4-6 referred to the TCC with zero applied volts (though you won't find a single vendor rubbing that fact in). Now look at Figure 4-7, which shows how the capacitance falls with applied voltage. As expected, you have 0% at 0V. But then hold your breath. The capacitance falls anywhere between 25% to 65% by the time you get to the rated voltage. Remember, I had assumed better voltage stability even in my previous "worst-case" estimates. This just goes to show, when it comes to ceramic capacitors, we can't assume anything. Pore over the vendor's datasheets very closely, keep asking those embarrassing questions, and ensure that your smart aleck purchase officer doesn't slip in another vendor when you are not looking.

Notice another interesting thing in Figure 4-7. There are two 10nF/50V capacitors in the 0603 size. The fall in capacitance of one is around 25%, the other 55%. Clearly, these two capacitors are very different as far *you* are concerned. But if you look at their part numbers, you may have to look for almost 20 minutes before you see any difference (I certainly had to look almost that long). I think there is some difference in the 10th or 12th digit of the part number (an additional zero somewhere). Don't you think that practice is virtually the same thing as manufacturers of semiconductor ICs tucking potentially embarrassing information on page 12 of their datasheets?

Now look at Figure 4-8. You see that it also gives a huge 45% fall in capacitance by the time you get to its maximum voltage (100V). But notice that if you use this capacitor only

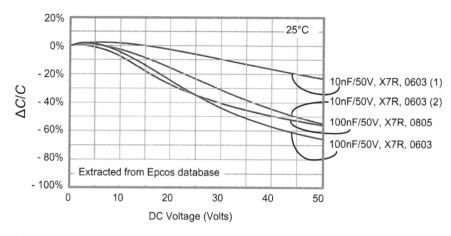

Figure 4-7 Voltage Coefficient for Different Sizes of 50V Capacitors from Epcos

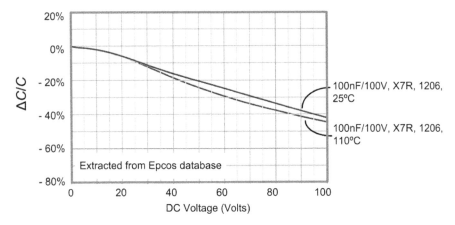

Figure 4-8 Voltage Coefficient for Different Sizes of 100V Capacitors from Epcos

up to 50V, it will give you much better voltage stability than a 50V capacitor. So, a 100V capacitor may be bigger, but it could be worth considering, even in a 50V application.

I have not implied that Epcos makes *bad MLCCs*. Most likely they are giving this parametric information out a little more honestly. But I tried running the Murata tool with capacitors similar to the Epcos capacitors, and found typically a 40 to 45% fall in capacitance by the time maximum rated voltage (50V) was reached. That is probably within the same ballpark (except of course for the fact that this particular ballpark is on the opposite side of town from where you thought you were headed).

I should also point out that 25V rated capacitors from Murata have a much better voltage stability curve—only 5 to 10% fall at maximum rated voltage. That is actually close to the voltage stability figure I used in my worst-case calculations previously.

ESR Dependence on Frequency for 100nF Murata Capacitors

In Figure 4-9 I have superimposed several curves using the Murata design tool. All are 100nF capacitors, the type I strongly recommended in Chapter 2 as the decoupling capacitor of choice at the input to a typical switcher. Note that Murata has a Low-ESL series that gives much lower ESR, too. One thing we can be certain about is that ESR falls with frequency, just as it does with elkos. However, at some point, the ESR starts to rise again. Note that this variation of ESR must refer to AC resistance effects. It has no plausible relationship to the resonant frequency or the ESR *at resonance*. For these 100nF capacitors, the ESR has a minimum between 5MHz to about 10MHz. Surprising at first sight, the generic 1206 capacitor seems to perform more poorly than the 0402 capacitor, because its ESR starts rising at a much lower frequency. We would have thought the larger package

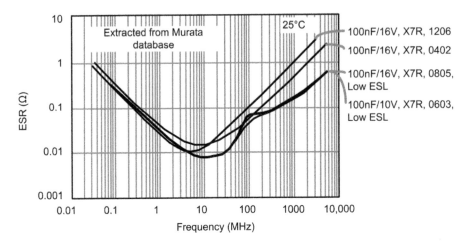

Figure 4-9 Comparing ESR Versus Frequency in Different Sizes for Murata's 0.1μF Ceramic Capacitors

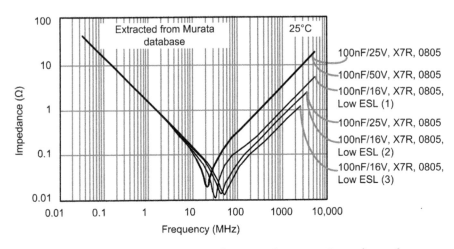

Figure 4-10 Comparison of Impedances of 100nF Capacitors from Murata

gave more opportunity to lower the interconnect resistances (DC and AC). On closer observation we see that we are partially right, because the lowest ESR achieved by the 1206 is certainly lower than the lowest ESR of the 0402 package. Unfortunately, as the frequency increases, the ESR of the 1206 gets substantially worse than the 0402.

The lowest point of ESR occurs at a much lower frequency than the resonant frequencies of these capacitors, as indicated in Figure 4-10. In the latter figure, we have several 100nF capacitors, all in 0805 size. We see that the resonant frequency ranges between 20 and

60MHz. Low-ESL capacitors perform much better, obviously. Some of the capacitors are the same as in Figure 4-9, and it becomes clear that the ESR reaches a minimum well below the resonant frequency of the capacitor.

So when somebody tells you that the capacitor is good to 30MHz, say it is better at 10MHz *in more ways than one*. Or if someone says the ESR of the capacitor is "only 30mΩ," tell him or her that at low frequencies the ESR is almost a hundred times greater (at least as per Murata's tool, but read the next section on ESR of Epcos' capacitors, too).

100nF Capacitors from Epcos Compared

In Figure 4-11, we see several 100nF capacitors from Epcos. Their resonant frequencies range from 20 to 40MHz. The ESR is also plotted out, and we see it does tend to bottom out slightly before its impedance curve. However, the Murata ESR curves and the Epcos ESR curves look very different otherwise. I certainly expect any impedance curve of a capacitor to fall at −20dB/decade (a factor of 10 for each decade in frequency), and then ultimately, when it becomes inductive, to rise with a slope of 20dB/decade. But I don't think that ESR needs to have a ±20dB/decade slope as seems indicated in Figure 4-9. AC resistance effects are a little more complicated than that, I would have thought.

I think it is important not to always presume vendors of components are infallible—neither the IC manufacturers nor the discrete suppliers. We must look at any data with our own judgment and experience. I personally think Murata needs to revalidate their ESR data, or at least the design tool that generated the curves shown in Figure 4-9. It really seems odd.

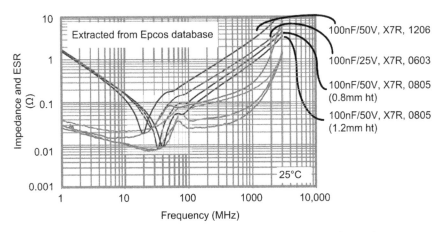

Figure 4-11 Comparing Epcos' 0.1μF Ceramic Capacitors in Different Sizes

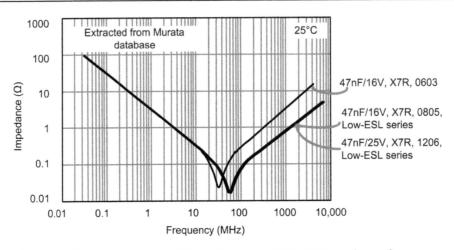

Figure 4-12 Comparison of Impedances of 47nF Capacitors from Murata

47nF Capacitors from Murata Compared

I was wondering that if 100nF were supposed to be such a golden value, would I fall off a cliff or something, if I reduced the capacitor to, say, 47nF? Nothing really changes dramatically as you can see from Figure 4-12. The resonant frequency is probably slightly higher in general, and the ESR slightly greater, but really, if there were the slightest difference in cost or size, I would opt for the 47nF.

Variation with Temperature

Though such capacitors usually lose their capacitance at high temperatures, their ESR typically improves. I picked a 10nF capacitor in Figure 4-13 and varied the temperature using Murata's design tool. The resonant frequency hardly changes, but the ESR does improve with temperature. That, like the previously reported dependence on frequency, is once again similar to an elko. But luckily, this does not cause the ceramic capacitor to evaporate as an elko would!

Recommended Derating for Ceramic Capacitors

You should remember that there is a maximum safe dissipation based on package size. That is an important consideration when you pass AC current through the capacitors. That dissipation depends on the particular manufacturer, but can be typically between 150mW and 200mW (for any size from 0402 upwards). Further, since there is an upper temperature limit, the allowed dissipation is derated, that is, linearly reduced above 40°C all the way down to 0mW at 125°C (for X7R).

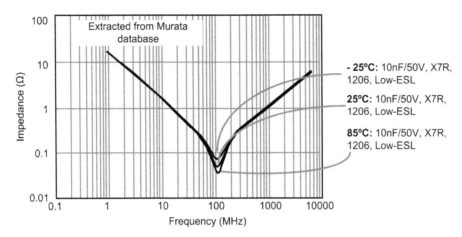

Figure 4-13 Variation of Impedance with Temperature for a 10nF/50V Capacitor from Murata

Note that ceramic capacitors have such low ESRs that it is not very practical to talk about the current passing through them. So manufacturers typically provide a derating based on the changing *voltage* waveform across them. Typically, if the rated DC voltage is V_R, then the peak-to-peak voltage of the waveform is not allowed to exceed 80% of V_R. In addition, its RMS value is not allowed to exceed 28% of V_R.

Then to account for the fact that the actual dissipation depends on ESR, which increases substantially at low temperatures, a derating on the allowed RMS voltage is introduced. So at +5°C, we need to derate the RMS voltage by −7%, at −15°C by −15%, at −35°C by −20%, and at −55°C by −30%. But you may need to talk to your vendor to get information specific to his or her part. Everybody has differing guidelines since the technologies are quite diverse.

Aging of Ceramic Capacitors

Capacitor aging describes the negative logarithmic capacitance change that takes place in ceramic capacitors over time. The crystalline structure for modern barium titanate based ceramics changes on passing through its Curie temperature (also known as the Curie point), which is about 125°C. This domain structure relaxes with time and in doing so, the dielectric constant reduces logarithmically. Further, the effect gets magnified the higher the dielectric constant is.

The aging process is reversible and repeatable. Whenever the capacitor is heated to a temperature above the Curie point, reset occurs and the aging process starts again from zero. Age reset can be formally assured by heating the capacitor for one hour at 125°C or for half an hour at 150°C.

The aging constant, or aging rate, is defined as the percentage loss of capacitance due to the aging process of the dielectric occurring over a decade of time (a tenfold increase in age). This means that in a capacitor with an aging rate of 1% per decade of time, the capacitance will decrease at a rate of:

- 1% between 0.1 and 1 hour

- 1% between 1 and 10 hours

- An additional 1% between the following 10 and 100 hours

- An additional 1% between the following 100 and 1000 hours

- An additional 1% between the following 1000 and 10,000 hours, and so on

That is a total of 4% after 1000 hours, *if the aging rate is 1%*. For X7R and X5R the aging rate is typically taken to be −2.5%, and for Y5V it is −7% (but check with your specific vendor). So that gives a 10% and 28% fall in capacitance for X7R and Y5V, respectively, at the end of 1000 hours.

Because of aging, it is necessary to specify an age for reference measurements at which the capacitance shall be within the prescribed tolerance. This is traditionally fixed at 1000 hours, since for practical purposes there is not much further loss of capacitance after this time.

Therefore, all capacitors shipped are within their specified tolerance at the standard reference age of 1000 hours, after having cooled through their Curie temperature.

There are apparently customers who soldered on ceramic capacitors in their power supplies and found the clock was just too low. They figured the capacitance was *above* the guaranteed upper tolerance band (a rare event with commercial ceramics!), and shipped them right back to their manufacturers. But the problem was only that as soon as the PCBs went through the soldering process, *age reset* (or de-aging) occurred and so capacitance rose. If only they had waited for some more time, their clocks would have been right on! However, I would have preferred SMD film capacitors if stability was so important.

ESL Is Important, Too

In modern processor applications, the allowed output tolerances of the converter providing the power rails have shrunk significantly. The core voltage requirement will typically be 1V with ±3% tolerance, including any AC transients, ripple, and DC accuracy! That's a maximum of ±30mV. In addition, very high dI/dt moments (current spikes) are demanded by the processor while operating. There is no single power management regime that suffices. What is required is a combination of very good point-of-load decoupling, fast control loops, high switching frequencies, droop techniques (i.e., dynamic voltage positioning), and so on.

The ESR of the decoupling capacitors has to be minimized, as does the ESL. As can be seen from Figure 4-14, *everything* becomes important. In effect, this is the final search for the ideal capacitor! Though there is no ideal capacitor, they certainly are getting better and better, as the next sections reveal.

Incidentally, why are droop techniques actually helping stay *within* the tolerance window? That sounds like a paradox, right? But take a look at Figure 4-15. You can see that after suddenly going to max load, if the DC value (settling value) is as low within the allowed DC window as possible, that actually gives more room (overhead) for the AC spike that will occur the moment the system slips into no-load condition again. Similarly, if after that event, the output voltage settles as high within the DC window as possible, there is more room for the downward spike that will occur the moment the processor suddenly demands max load again. In other words, you are dynamically positioning the DC level to eke out the maximum advantage that the full AC+DC window allows for the given processor.

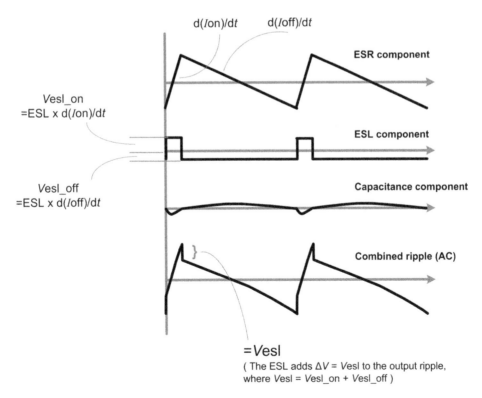

Figure 4-14 Contributions of the Output Capacitor to the Observed Voltage Ripple in a Buck Converter

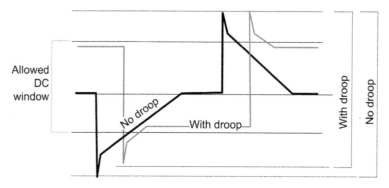

Figure 4-15 Droop Method (Dynamic Voltage Positioning) Allows More Room for AC Transients

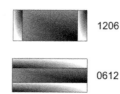

Figure 4-16 Reverse Geometry Capacitors for Lower Inductance

Reverse Geometry MLCCs

We were used to the 0805s, the 1206s, and so on. But manufacturers have realized that using *reverse geometry* helps significantly lower the ESL of their standard capacitors. They have several low-inductance capacitors in this new form factor (see Figure 4-16). Note that 1206 was of length 120mil and width 60mil. No! The 0612 is not of *length* 60mil and *width* 120mil! That does not change, only the terminations have shifted from being *along the width* to being *along the length*. It is like watching a 4:3 movie and a 16:9 movie on your new laptop's screen. In one case you have bars on either side, in the other case the bars are above and below. But neither seems to fit your screen comfortably!

Extra Low Inductance Capacitors

Many vendors have come up with interdigited capacitors (or IDCs). These extra low ESL capacitors are based on the same principle with which I built my monolithic 5V/50A Flyback described in Chapter 5, titled "Maximizing the Effectiveness of the Ground Plane." See Figure 4-17. As you can see, they must have been a pain to lay out inside the chip, but wait until you try to connect them on your PCB without defeating the very purpose they

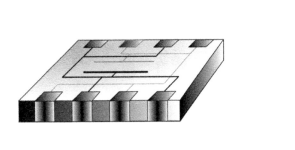 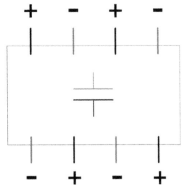

Figure 4-17 Interdigited Ceramic Capacitors for Low Inductance

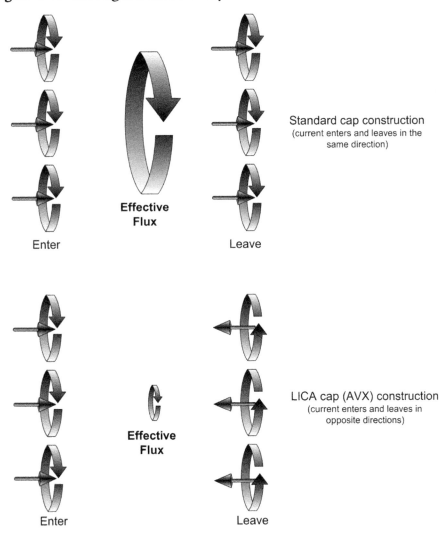

Standard cap construction
(current enters and leaves in the
same direction)

Effective
Flux

Enter

Leave

LICA cap (AVX) construction
(current enters and leaves in
opposite directions)

Effective
Flux

Enter

Leave

Figure 4-18 Principle of the Low Inductance Chip Array (LICA from AVX)

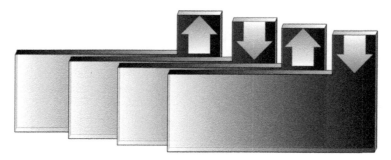

Figure 4-19 How the Current Paths Inside the LICA (from AVX) Minimize Inductance Further

were made for—low inductance. With the positive and negative traces having to weave in and out, you will have no option but to use a bunch of vias every now and then. Maybe they just managed to transfer a good part of the inductance inside the capacitor onto your PCB!

Similarly, AVX has come up with improved IDC capacitors called LICA capacitors (low inductance chip array). They were developed in a joint effort between AVX and IBM. Their basic principle also remains the same—flux cancellation by opposite current flows. (See Figure 4-18.) They look and feel like regular IDCs (and need to be laid out similarly), but they have an improved internal electrode structure to further minimize ESL. See how the currents are forced inside the chip in Figure 4-19.

The gains for the popular 100nF decoupling capacitor in a package of about 80mil × 50mil (0805 or similar) are summarized as follows:

a) Standard geometry capacitor: Resonant frequency 30MHz, ESL 800pH

b) Reverse geometry (conventional): Resonant frequency 55MHz, ESL 130pH

c) Reverse geometry low-ESL (IDC): Resonant frequency 80MHz, ESL 50pH

d) Reverse geometry LICA (IDC): Resonant frequency 100MHz, ESL 25pH

Maximizing the Effectiveness of the Ground Plane

How to Parallel Output Capacitors for Proper Sharing

This goes back to my days at a well-known manufacturer of three-terminal, high-voltage switcher devices. To the uninitiated, these devices may look like a TO-220 Mosfet. But they are really quite remarkable, with an integrated 700V Fet and control, all on one die. One particularly nice thing about these devices is that the tab (of the TO-220 package) is at ground potential (connected to the Source terminal). So you can connect a heatsink to it without any insulation, and the heatsink won't end up spewing electric fields everywhere. The alternative is to use a controller IC driving an external Fet, for which you would need to insulate the tab of the Fet from the heatsink (the Mosfet tab is usually connected to the Drain, which is a swinging voltage node in a Flyback application). Then you would need to ground the heatsink to prevent it from becoming an antenna. Though it seems the part itself switches so fast, you do get a huge slew of EMI out of it anyway! Once in our 800W server power supply, the highest, most troublesome, and stubborn part of the EMI spectrum was actually discovered to be from the measly 25W standby power supply. This one was made from this particular monolithic integrated switcher!

However, it is still a nice device overall, and I would have thought you would never need to do the slightest "specmanship" to sell it. But ironically, I did learn much of that art right there in that company! Culture really proliferates, it seems. Because that company was created by a bunch of unhappy individuals who had spent several years in a major analog company (one that I worked for later). But thankfully, I also learned a lot about the Flyback topology itself, and really understood the importance of PCB routing, ground bounce, ground planes, and other basic topics. I am therefore convinced that the Flyback is really the most demanding topology in terms of PCB layout.

Let us take a look at Table 5-1. This is extracted from the relevant application note of the high-voltage switcher family (one that I incidentally wrote, under a fair amount of superior "guidance"). If you look closely, you will see something quite surprising. For example, in the last column (the 249Y), the required ESR of the (aluminum electrolytic) output capacitor is 1mΩ. Considering that even modern ceramic capacitors are said to present a very low ESR of 10 to 30mΩ, that value is clearly impracticable.

Table 5-1 The Parameters Used to Generate the Published Efficiency and Maximum Load Curves of a High-voltage Switcher Family

Typical 5 V output power supply component parameters Universal input (85-265 VAC)									
Parameter	Units	242Y	243Y	244Y	245Y	246Y	247Y	248Y	249Y
Current limit (typ)	A	0.45	0.90	1.35	1.8	2.7	3.6	4.5	5.4
Maximum transformer primary inductance Lp	μH	2780	1385	923	693	462	346	277	231
Transformer leakage	%/Lp	1.5	1.5	1.5	1.5	1.5	1.5	1.5	1.5
Inductance secondary trace	nH	20	20	20	20	19	16	13	10
Transformer resonant frequency (secondary open)	kHz	750	800	850	900	950	1000	1050	1100
Transformer primary AC resistance	mΩ	2000	1060	700	600	500	300	200	100
Transformer secondary AC resistance	mΩ	12	6	4	3	2	1	0.75	0.5
Output capacitor equivalent series resistance @100kHz	mΩ	18	9	6	5	4	3	2	1
Output inductor DC resistance	mΩ	6	4.5	3.5	3	2.5	2	1.5	1
Common mode inductor DC resistance (both legs)	mΩ	370	340	310	280	250	220	190	160
Core loss	%/P_{IN}	2	2	2	2	2	2	2	2
Current limit (typ)	A	0.45	0.90	1.35	1.8	2.7	3.6	4.5	5.4

We all know that the losses in the output capacitor of any Flyback are high, due to the choppy waveform of the current they encounter coming through the diode. It is obvious that reducing the ESR to zero would totally knock off a major chunk of losses, boost the published efficiency curves, *and* allow a much higher maximum achievable power for the device. However, "0mΩ" would have been much too obvious, wouldn't it? So it was a case of "Buck the ESR and Boost the efficiency." In other words, a perfect Buck-Boost.

My problem was that the company now expected me to somehow "validate the efficiency curves." They were perhaps overly sensitive about possible legal complications, such as, if the customer demands to see a working board to reaffirm the datasheet, and there happens to be none. Quite like stock options with no dates to back them up.

So after having created the math spreadsheets and having "improved" the efficiency curves, I actually had to do the impossible now. I had to *build* it. What worried me was not just the fact that I had to obviously parallel over 10 large capacitors to get an effective ESR close to the target, but the fact that I had to ensure that the impedance of the intervening PCB traces also virtually canceled out.

On a typical one-sided board (still very common in commercial AC-DC power supplies), *as the number of capacitors you try to parallel goes up, so does the intervening PCB trace impedance*. Take, for example, Figure 5-1, where we have the simple case of one output capacitor. A small advisory here—if you try to reduce the impedance further by making the current loop smaller and smaller, the capacitor would eventually start comparing notes with the heatsink on the topic of temperature, and that can't be good for its life expectancy. There is also a major issue concerning *secondary-side* trace inductances, one that we will discuss a little later. Other than that, there are no issues, except of course the fact that because there is only one capacitor, the effective ESR won't be very good (nor the RMS ripple current-handing capability).

So suppose we try to parallel three similar capacitors, as shown on the *left side* of Figure 5-2. By following the current paths, we realize that the outer capacitors are going to be less and less effective in terms of sharing current. Therefore, we need to provide a little ballasting. A horizontal cut is made in the lower island as shown in the schematic on the *right side*. This ensures that the total PCB length, as seen by the currents, is identical for all

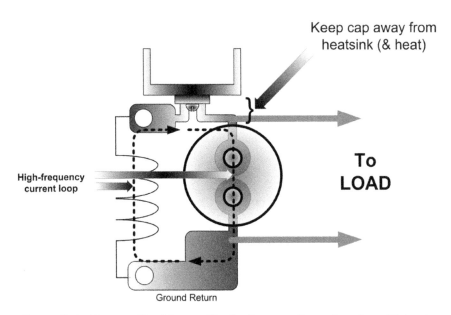

Figure 5-1 How to Position a Single Output Capacitor in a Flyback

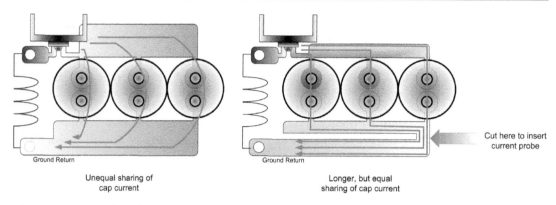

Cut here to insert
current probe

Ground Return

Ground Return

Unequal sharing of
cap current

Longer, but equal
sharing of cap current

Figure 5-2 Good and Bad Ways to Get the Output Capacitors to Share the Current in a Flyback

three capacitors. Further, if we ever want to measure the RMS current through the output capacitors, we need to make a cut at the location indicated and insert a current probe. We can then say with confidence that each capacitor will be carrying one-third of whatever current waveform we observe at this point. But we can't do the same by just adding capacitors blindly on the PCB.

I remember in a previous company making AC-DC power supplies, whenever their meticulous design integrity group would insert a small loop of wire in series with *one* of several paralleled output capacitors, to insert a current probe and measure the RMS current through it, the inductance from that little wire would itself manage to *divert* the current away from the capacitor being measured, into the other capacitors. There is really no way to measure the RMS current through each capacitor individually with any degree of confidence. So the best bet is to take pains to first ensure equal sharing as indicated above, and then record the waveform of the current through *all* the capacitors at the same time, and then divide that waveform by three (scale it down in a vertical direction). Note that a waveform one-third the size of the original has its RMS reduced by a third too. That is why, for example, if we need a ripple current handling capability of 12A, we need four capacitors of 3A each, or three capacitors of 4A each, and so forth, *but they must be well paralleled.*

What if we place a small sense resistor instead of a current loop, to read the current through a capacitor? Well, if we use a very small sense resistor, we will have so much relative noise, our readings would be unreliable. If we use a larger sense resistor, we would still divert current into other paralleled capacitors. Again, the correct solution is the same as for the current loop, measure *all together*, and then divide by the number of capacitors.

The above reasoning convinced me that my task seemed completely hopeless. There was no way I could get nine or ten capacitors to share current well without increasing the trace

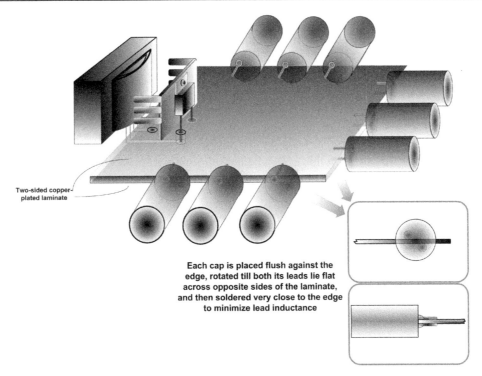

Two-sided copper-plated laminate

Each cap is placed flush against the
edge, rotated till both its leads lie flat
across opposite sides of the laminate,
and then soldered very close to the edge
to minimize lead inductance

**Figure 5-3 How a 5V/50A Flyback Prototype, Running off 85VAC, Was
(Barely) Realized**

impedances beyond imagination. But I finally did solve the problem *with a two-sided board*,
and thus created a magnificently impractical 250W Flyback prototype, delivering 5V @
50A from 85VAC input. Its measured efficiency was around 62%, in line with the published
datasheet curves, and also the predictions of my Mathcad spreadsheet (that spreadsheet is,
incidentally, available on the accompanying CD-ROM of my "A to Z" book). The output
stage of my grizzly creation is shown in Figure 5-3 (try selling this one!). But it works!

Note that I was not willing to allow even a millimeter of (external) lead inductance to come
in the way of my grand designs. And by paralleling two large 2-oz copper planes carrying
forward and return currents, I *virtually canceled out all intervening impedance*. I was left
with only several capacitor ESRs, all in parallel, nothing more. Of course there was no
way I could *demonstrate directly* that indeed all intervening impedances had been
virtually canceled. The proof of this pudding was only in the eating (or in the heating, in
this case).

It is important to keep in mind that the Mathcad spreadsheet I mentioned had by then been
very successfully validated (in excruciating detail) over several of their previous product
families, and was consistently giving agreement with bench results to within ±1% accuracy

over all line and load conditions. Otherwise I, for one, would certainly have brought my own spreadsheet into question first. Incidentally, the device does have a significant amount of crossover loss (about 30 to 40% of the total switch loss at high line, despite the fact that their still-available application note AN-19 states that it is negligible).

Integrated Switcher IC Solutions Versus Controller IC Solutions

Of course the only way to keep this 250W behemoth running for more than a minute without hitting the IC's internal OTP (over temperature protection) was to use a water-cooled heatsink, which my rather lab-ratty senior tech fashioned in less than half a day with some tubes, strips of metal, high-temp solder, and a sump (submersible pump). Note that the dissipation in the TO-220 was about 15W at this power level (calculated and measured). Yes, if you think that is a wee bit too much for a TO-220 to handle, I am on your side.

You must have also realized that the advertised max load power rating of this entire family of devices has nothing to do with its max junction temperature (or the size of the heatsink required). Their load/power rating is based purely on the criterion of not hitting the internal switch current. The Fet junction temperature is assumed to be 100°C, and *steady* in that calculation. How to keep it there was considered your problem entirely. "Yes we know, the Fet has a drop of 18V across it at its max rated peak current, and damn right it gets hot, but do you think we are in the business of selling heatsinks?" Note that this is completely unlike the method used to ascribe power-handling capability to commercial Mosfets. So if you think you can happily migrate from an application with a 6A Mosfet to one with a "6A" monolithic switcher, you will be in for a surprise. You may discover that the so-called equivalent switcher has an Rds (Drain to Source on-resistance) typically 2 to 3 times higher than your expectations. I have thus learned never to compare apples to oranges (even if all the apples in the world suddenly start looking like oranges one fine day, thanks to genetic engineering or clever marketing). So I agree this may have *looked* like a TO-220 Mosfet to me once upon a time, but then, in those days I was still young.

In a later section we will understand more clearly why the output stage as shown in Figure 5-3 even worked.

Quick Check on Current through Aluminum Capacitors

One quick test of whether the current passing through an electrolytic capacitor is within bounds is to *touch it* after it has been running for some time (high-voltage power supplies must be turned OFF just prior to this!). If an electrolytic capacitor has been designed with the normal recommended procedures for ensuring its life, the delta between its case temperature and the ambient temperature should be almost equal to the delta between its case and its core. Most 85°C capacitors are designed for about a 10°C differential between core and case (when passing their rated ripple current), whereas most 105°C (long-life) capacitors are designed for a 5°C to 8°C differential. So the case of any electrolytic capacitor should never exceed 35 to 45°C, at a room ambient temperature of 25°C. It should be reasonably warm to your touch, not much more, otherwise you probably have a ripple current cum life expectancy problem. Read Chapter 4, titled "Using Capacitors Wisely."

Secondary Side Trace Inductances and Their Impact on Efficiency

Many years ago, after having developed some initial confidence making my very first 70W Flyback in Bombay, I thought to myself, "why can't I go all the way? Why can't I, say, make a *600W Flyback?*" It is like if you don't know about the existence of gravity, you can fly. So I *did* make that 600W Flyback, and also later wondered what the fuss was all about. *Flybacks rocked!*

Incidentally, power Mosfets had just started becoming commercially available in my town, and the market price was a couple of dollars each. My colleague was happily chomping his way through them, almost at the rate of 5 to 10 a day for his pet inverter project. They were relatively fragile and failed easily, if for example the *reapplied* dV/dt was high. So instead of taking that risk, I settled for several paralleled TO-3 bipolar transistors (BU-208s if I remember), with ballasting resistors in the base and emitter of each, all mounted on a large metal chassis. For the magnetics, I stacked two EE65 cores side by side and managed to put a hand-wound coil around the stacked center limb. The output capacitors were not too many or too big either, and the overall efficiency was an impressive 70% or so. The question that haunted me, but only *many years later* was—how did I ever do *that?*

The answer to that has *three* parts to it. First, I was only using the 230VAC available from the wall, and so the input current was much lower. Second, I had fortuitously happened to choose the output as 60V @ 10A, maybe because I was secretly hoping I could use it eventually to power a Class AB high-power audio amplifier. The output currents were therefore much lower than if I had decided on, say, 5V @ 120A. That is why the number of output capacitors, their ESR, and the ripple current capability, were all pretty manageable. Third, what also helped, unknown to me at that time, *was that the turns ratio of my transformer was low*, about 5:1 or 4:1 if I remember. Turns ratio happens to be the single biggest reason why the efficiency of a typical Flyback plummets at high loads. *If you are haunted by poor efficiency in your offline Flyback, and your output voltage is around 5V or less, this is where you need to start looking.* Remember, a 5V output has a typical turns ratio of around 20:1. Check the temperature of your zener or RCD clamp. If it is sizzling beyond your initial estimates, the leakage is much higher than you thought. And the reason may not be your finicky transformer vendor, but *your own PCB layout.*

Here is how I finally learned the truth about turns ratio. In this high-voltage, three-terminal, monolithic switcher company, while validating my Mathcad spreadsheet (in its early stages) against their previous products, I had managed to reach very good agreement with bench results for all boards set for 12V output. But for boards set to 5V, at higher loads, my efficiency predictions were at least 5 to 10% too optimistic than observed reality. I just couldn't understand that for weeks on end. My spreadsheet was by then in complete agreement with the predictions of their own internal Excel spreadsheet, the one that had formed the basis of all their previously published efficiency curves. Finally, after desperately plunging into a prolonged literature survey, I realized the reason for the discrepancy. *All secondary trace inductances in the high-frequency path reflect on to the primary side, multiplied by the square of the turns ratio*, and numerically add on to the existing primary-side leakage inductance to give the total effective leakage inductance as seen by the switch (and by its associated zener/RCD clamp). For example, a typical 5V output Flyback would have a transformer turns ratio of about 20 to 25 (more than that would damage the 600 or 700V Mosfet because of the reflected output voltage, and too low a turns ratio for a 5V output would imply a very low duty cycle and poor efficiency). So even an inch of PCB trace (20nH estimated) will reflect on to the primary side as $20^2 \times 20\text{nH} = 8\mu\text{H}$. That is very significant if you consider the fact that a typical well-designed transformer has a (primary-side) leakage inductance of about 1 to 2% of the inductance of the primary winding, and the first estimate of the zener/RCD clamp losses can still be a sizeable portion of the total losses. So for a typical 1mH universal-input 70W Flyback transformer, for example, the primary-side-based leakage is about 10 to 20μH, *which almost doubles, just because of one inch of uncoupled secondary-side trace inductance*! The dissipation in the clamp also almost doubles and the overall efficiency plummets. That is one of the most important reasons for trying to minimize that current loop area indicated in Figure 5-1. It is *not just EMI*, but efficiency that is affected. The full effect of PCB trace impedances can never be more dramatic than that.

You can argue—but 20nH per inch is just a rule of thumb! How can we even confirm what the effective leakage inductance really increases on the primary side (as a result of that)?

The oft-suggested method of measuring leakage is to short the secondary pins and measure the inductance across the primary winding. That is shown in the upper schematic in Figure 5-4. But we now realize that may be OK for spot-checks to confirm the quality of the transformer in production, it just isn't good enough for the rest of the circuit. What we need is an *in-circuit* measurement of the effective leakage. The method I finally suggested to the high-voltage switcher company is indicated in the lower schematic of Figure 5-4. If you do this, and compare it to the value you got from the upper schematic of Figure 5-4, you could be in for a real nasty surprise.

Finally, after all their internal checks and balance (the company was indeed fastidious in matters of product liability), they agreed they had gotten to understand the Flyback topology far better now. So I quickly incorporated the new loss term into the Mathcad and Excel

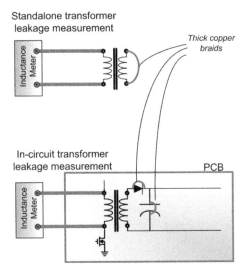

Figure 5-4 How to, and How Not to, Measure Effective Primary Side Leakage Inductance in Flyback

spreadsheets, and the fit was (almost suspiciously) perfect! Of course we decided it was not "practical" to revisit the previously published efficiency curves and fix them, even though the results were finally in line with what customers had been reporting all along for the previous families. No point alarming customers by telling them they may be in trouble.

Where can we find thick copper braids? Well, I tend to use regular solder wick lying around the lab. I am fairly convinced by now that if I partially soak some solder into it, it works great as a high-frequency shunt. I won't be caught ordering Litz wire just yet.

Current Return Paths in the Ground Plane

Let us try to understand why the prototype in Figure 5-3 ever worked. In Figure 5-5, we have a right-angled trace carrying current on the top side of a 2-oz copper PCB. If the current is DC, it chooses the path of least *resistance*, which is a straight line through the ground plane as shown in the upper diagram. If the frequency is very high, the return current chooses the path of least *inductance*, which is in parallel with the upper trace, as shown in the lowermost diagram. For mid-frequencies, the return current follows an intermediate path. In all cases though, the current is essentially trying to choose the *easiest* path (*lowest impedance*), the difference being that at low frequencies, the impedance of any conductor is predominantly resistive, whereas at high frequencies it is mainly inductive. So at high frequencies the return current tries to minimize the enclosed area shown in

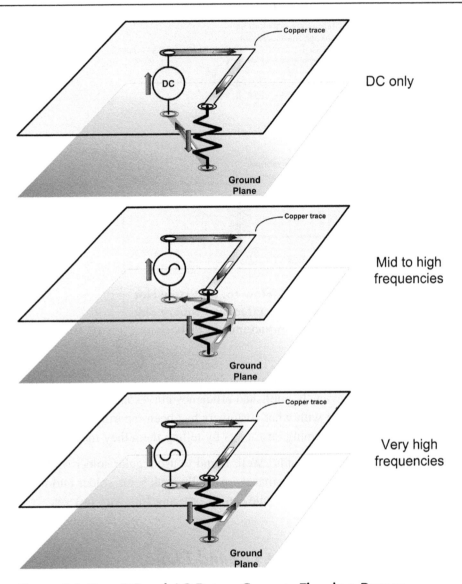

Figure 5-5 How DC and AC Return Currents Flow in a Proper Ground Plane

Figure 5-6, and tries to couple very closely with the forward trace. In doing so, it attempts to cancel the fields produced by the forward trace. It is nature's way of helping us, if we just let it. We should realize that inductance exists only because of the field it produces and vice versa. The field contains the associated stored energy of $1/2 \times LI^2$. Then if the field is somehow canceled, so is the inductance. Proximity effects are also minimized by the paralleling of these opposite traces, and therefore AC resistance is also reduced.

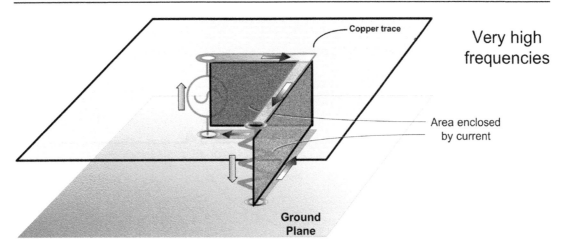

Figure 5-6 The Area that Is Sought to Be Minimized by the High-frequency Current Components

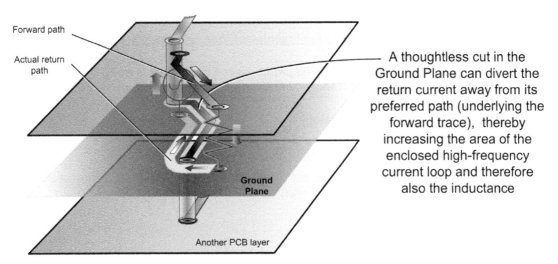

Figure 5-7 How Cuts in the Ground Plane Can Reduce Its Effectiveness

You should be very clear that under the influence of external fields, the obvious solution may be the most deceptive. A straight line is not so straight (or "short") any more! See the legendary Brachistochrone Problem in Figure 5-8. I had actually built this for a science project in my Physics days and, more recently, for my daughter's prize-winning science project.

Finally, in Figure 5-7 we understand how a thoughtless cut in the ground plane (perhaps for routing between adjacent layers) can prevent the current from flowing in its desired path and effectively ruins its effectiveness. Further, we have also managed to create a *slot*

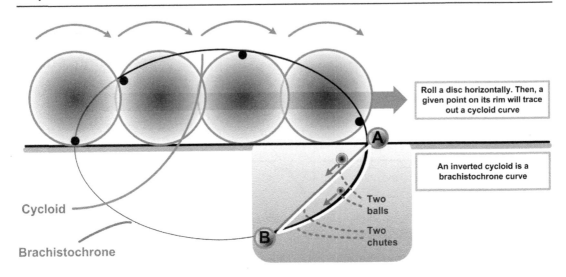

Two balls (of any size) rolling down between two points A and B:

The ball rolling down the brachistochrone path always wins

In general, the preferred path between two points is not necessarily the shortest in length (i.e., a straight line)

Figure 5-8 The Brachistochrone Analogy

antenna on our PCB. Eventually, we can get circulating currents flowing in a loop around such a slot, emitting H-fields and causing EMI.

Paralleling Traces to Reduce Inductance

A major Taiwanese customer was once having trouble with the DC-DC power converters on their rather complex board meant for a high-end notebook computer. My company eventually ended up with a rather bloody nose, losing several million dollars in a derailed deal. But it was not really all their fault. When I looked at the board, I was quite aghast.

One of the things the customer had done was use 1μF decoupling capacitors, whereas we had clearly recommended 0.1μF capacitors. A 1μF ceramic capacitor has a resonant frequency of around 10MHz, whereas a 0.1μF capacitor works up to 30MHz. When I asked why they had used 1μF capacitors, secretly praying and hoping they had a good reason for it, all their engineer could come up with was "1μF? Better than 0.1μF, no?" (In America, isn't bigger supposed to be better?) That answer didn't do much to enhance my confidence in their engineering capabilities (at least vis-à-vis Power Conversion), though it didn't really turn out to be the main issue either (*at least not on the few boards I examined*). Their Fets were blowing up unpredictably, and I could see at least one major reason for that. I think maybe they had concluded that our controller IC was for some reason completely allergic to n-channel enhancement-mode metal-oxide semiconductor field-effect transistors—because these were at least three inches away, connected through traces routed as carelessly as it gets. What's *so* wrong about that you may ask!

Well, for one, the Gate drivers need to momentarily push in close to 1A through the Gate traces to get the Fets to switch quickly and efficiently. Three inches of Gate impedance cannot be too good for that. Second, one of the things you need to remember about Synchronous Bucks is that they have a builtin deadtime to avoid cross-conduction through the Fets. That interval is typically about 10 to 20ns. So if you introduce any *asymmetry* between the drives, you could easily end up effectively nullifying the deadtime, causing cross-conduction. Look at Figure 5-9 to see what was probably making the Gate drive waveforms so lopsided (as seen *at the Gates*).

You have to follow the gray current loops that show how the Gate drive current flows. The lower Fet is *in effect* "very close" to the IC, because the ground plane cancels most of the

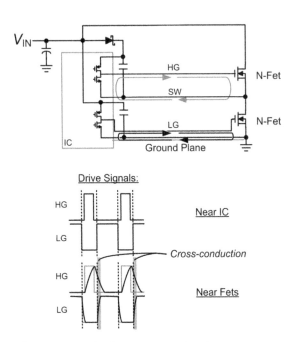

Figure 5-9 Placing N-Fets Far Away on a Two-sided Board Can Cause Cross-conduction

impedance of its intervening Gate trace. Remember, this was also the reason why that 5V @ 50A behemoth Flyback discussed previously ever worked. But the upper Fet Gate drive current returns through the SW trace, not the ground plane, thus *not* being bestowed by the same generous amount of impedance cancellation. So this particular Fet is, electrically speaking, far away. And that delays the signal arriving at its Gate relative to the other Gate drive, negating the deadtime and causing cross-conduction.

Note that if the upper Fet were a P-Fet, its driver's return current would usually flow through the ground, not the SW trace (there would be no bootstrap circuit present). So a P-Fet/N-Fet combination is actually more forgiving in terms of PCB layout (so long as we have a good ground plane!).

Note that there are people trying to make controller ICs with internal high-side N-Fet drivers that are not floating, but referenced to ground. These can be spotted by the fact that they have no Pin marked SW. If there is a SW node connection, it may just be there for current sensing. So no high-side driver return current passes through the SW node. You may think that this would be a good idea in terms of canceling trace inductance of the upper Fet, too. But unfortunately these drivers have a serious conceptual problem; at best the high-side gate drive signal can go to ground. But remember that in a Synchronous Buck, the SW node can go several 100mV or more *below* ground when the current freewheels. And in fact during the preceding deadtime, it could possibly be about 1.5 to 2V below ground (assuming no Schottky is placed across the low-side Fet). But the SW node is also the Source terminal of the high-side Fet (whose Gate is being held at ground). In other words, the V_{GS} during the deadtime can be about +1.5V, and can therefore turn the high-side Fet on momentarily, causing shoot-through, and loss of efficiency. So *"ground-referenced" high-side N-Fet drivers should be avoided*! Check your controller IC once again! An example of such a device is the 2743. In fact the datasheet of this particular IC forgets to ask you to connect a Schottky for any of the reasons I pointed out above, and neither does their eval board suggest any need for this rather expensive external component. Take your chances!

Multilayer Boards and the Ground Plane

To maximize the effectiveness of the ground plane, ensure it is the layer just below the component layer. That brings it as close to the power traces as possible. Not only does that help it couple well magnetically to the corresponding high-frequency traces above it, the increased capacitive coupling also helps sink some of the noise into the ground plane. If the ground plane is "good" (preferably 2-oz copper and very few cuts), it will continue to behave as a *quiet* ground plane. But too much injected noise can disturb its sense of balance. And thereafter, yours!

Deadtime in Synchronous Controllers and Switcher ICs

I was once visiting the Colorado design center of my analog semiconductor company. I had thought their IC designers were really talented. That admiration got critically damped when I heard them saying their new controller IC had no deadtime at all. They explained their new drivers switched very fast (within 1 to 2ns), and therefore they saw no need for any deadtime. Of course designers always seem to have a bunch of simulations to prove anything. Well, maybe the drivers themselves didn't need any deadtime because there truly was no chance of cross-conduction through *them*, but what about the external Fets they were driving? My experience was that in general, you can manage with only 10 to 20ns deadtime for *switcher* ICs (with integrated Fets), but for *controller* ICs, which could be driving Fets quite far away, you need to leave at least 40ns or so. You never know how the end customer may have routed the Gate drive traces, or the characteristics of the Fets they would eventually use. Never forget the fact that power Mosfets, however fast they may be, have an internal delay before they even respond to a pulse arriving at their gates. That delay is simply the time for the Gate-to-Source capacitance to charge up from zero to the threshold voltage of the Fet (assuming a perfect square pulse applied at the gate). Further, since only rarely do we use the same Fet for the high-side as for the low-side, their respective delays are also hardly likely to be the same. In other words, they now have inherent *relative* delay. All these effects can create cross-conduction, especially when combined with any inherent inadequacy of the deadtime being provided by the driver stages.

Equally jitter-inducing was the fact that the Colorado design center IC was to be a *current-mode* controller. In current mode, you invariably need leading edge blanking to quench the current spike at turn-on. Otherwise the comparator would respond to that noise spike rather than the rest of the smoothly rising inductor current sensed waveform. But these guys also proudly declared they had no blanking time whatsoever. "Not necessary at all" was what they said with a hint of bravado. But my experience again is that the amount of leading edge blanking required is roughly the same as the deadtime, and with the same criteria, too. In most high-voltage 3842-based applications, I actually had to have at least 100ns of blanking to get it to work right.

So I came home and helpfully prepared a lengthy spreadsheet to account for PCB trace impedances, Mosfet delays, and so on, as well as a survey of competitors' ICs listing their deadtimes and blanking times. But I couldn't get past their impregnable wall of simulations. So later when I heard their new silicon had had "some problems" I had a strange inkling! Apparently, their dream project went on for at least a year more, sucking in resources (mainly more simulation time I guess). I am not even sure they ever released it. Because for me, it was by then just a case of "don't ask, don't tell."

Printed Circuit Board Layout for AC-DC and DC-DC Converters

Introduction

The surest recipe for disaster is to give your schematic to a typical CAD person and walk away to "do more important tasks." Nothing is more important than PCB layout. It is the bridge, the *only* bridge, connecting your wonderful Mathcad-verified, Spice-validated, IEEE-published, Patent pending paper design, and a trouble-free finished *product*. So just make sure it is not a Half-Bridge you're unintentionally making!

You know by now that schematics lie. So how could a typical CAD person know how to interpret your schematic correctly? He or she could end up laying out a Buck just the same way as a Boost or a Buck-Boost. What's the difference, huh? The truth is every topology is quite different in terms of its recommended layout. So, when I take a look at a troubled Buck converter, I first look for assurance about something very specific in its layout. When I look at another topology, there is something else that quickly tells me what could possibly be wrong with it, in terms of its layout. These are the type of things I will try to clarify here. But there is a something else to talk about before we get there—the concept of eval boards!

Evaluation Boards (EVBs)

In every semiconductor company I have worked in, we have been arguing about this since time immemorial. As of last notice, no one has yet figured it out completely. The discussion goes typically as follows (just hang in there). Should we make an eval board only for the Apps guy to evaluate? But we don't have time, so should we also somehow make that board good enough to send to the customer? No, no, the customer needs to see a very *small* board, this won't do. Do you think the Apps guy can use the same tiny customer's board for his own evaluation? Of course not. But I think the big board is good enough to send to the customer at least *for now*, since he may want to stick some probes onto it and see the waveforms. No, that's out of question, we *don't* want them to look that so closely. But if we send a small board to the customer that has not been fully evaluated by the Apps guy, are you sure it will work properly? Oh by the way, the product engineer (PE) also wants a few eval boards to test. And so does the test engineer (TE). Can we somehow ensure this eval board can meet their requirements too? Shouldn't we make just one (grand) universal board? You know we also need to put it on that new automatic startup tester. OK, sorry, let's just make *three* different types of boards: one for the customer, oh sorry, *two* for the customer, one big and one small, then one for the Apps guy, and another for the PE/TE? That makes four by the way. Do you think there will be *correlation issues* between so many boards? Don't you really think we should make just *one* board for everyone? Oh by the way, the Web team is already selling some funny-looking "Build-It-Yourself" boards with big square copper islands. Did they even bother to send it to Apps to make sure it works? So do you think we can just use *that* instead? In any case, customers have been getting it, whether they like it or not. Stop! Are we talking about a *demo board* here, or an *eval board*? You *know* the difference right? And don't forget there's also a *customer's* eval board and an *Apps* eval board. These are *different* issues altogether! (Profound silence). So who is going to make them? We gotta make at least four boards. Oh by the way, maybe another one for the IC designer to validate too. But that one needs *sockets*, just like the one for the PE. What do you mean the part won't work properly with a socket? Is it some problem with the part? How about just for a simple functional power-up test? Isn't that what the PE wants? And so on and so forth!

The bottom line is you are likely to be troubleshooting (or building) a *whole bunch of* boards in response to everyone's demands. The boards will just come flying in, in a wide variety of PCB layouts. It can get very confusing if you forget that there are some *key characteristics they all need to share to guarantee proper performance.*

Buck PCBs

In Figure 6-1, we have three pictures of a Buck. The uppermost one shows all the traces carrying current when the switch is ON (shaded). The middle one shows what happens when the switch is OFF. The lowermost one is the *difference* of the two above; that is, it shows all the trace sections that *changed* in going from switch ON to switch OFF, thus indicating they are directly involved in a *transition*. In other words, these traces either had to *suddenly* start carrying current at a switch transition, or had to stop equally suddenly. These are called the *critical* traces (or *AC-traces*), the ones really carrying the high-frequency current harmonics. We need to pay very close attention to them. Because, as explained earlier, not only is their impedance to switch transition frequencies very high (producing ringing and noise), but they also develop tiny (but lethal) voltage spikes per $V = LdI/dt$. These can infiltrate deep into the IC, causing general control malfunction and chaotic behavior. Note that we are ignoring the pure DC component into and out of the board in Figure 6-1, and therefore also assuming perfect input and output decoupling/bypassing.

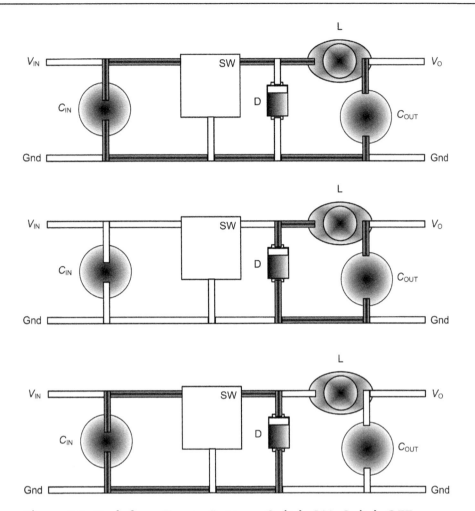

Figure 6-1 Buck from Top to Bottom—Switch ON, Switch OFF, Critical Traces

In a Buck we see that it is very important to have good input decoupling, and also to minimize the trace lengths between the ceramic capacitor and the IC. We have discussed that issue previously. But we now see that it is very important to also *minimize the length of the trace section connecting the SW pin to the common node of L and D*. This is one of the first things I try to check out when I troubleshoot a Buck switcher board handed over to me.

In a particular case I handled, a major Japanese customer was managing to blow up his simple Buck switcher under short-circuits on the output. I knew that these parts had not only the usual cycle-by-cycle current limiting (their first line of defense), but in fact, a hidden second level of current limit protection, which if ever encountered, caused protective foldback (skipped ON-pulses). Usually, it is almost impossible to go past even the first level of protection, let alone the second. So this was really mystifying. I actually set up two oscilloscopes, both on single acquisition mode, to see the event both *up close* (one scope set on 10µs/div) and also *zoomed out* (second scope on 10ms/div). They had identical triggering thresholds set, to capture the rising edge of inductor current. What I saw was really interesting. Once I shorted the output, the IC hit the foldback current level, and started skipping pulses as expected. The current decreased and so did the output voltage. Then at some point, the current started to build up again (attempted hiccup). But if you were expecting the IC to repeat its earlier stellar performance, you were in for a surprise. Because the *second* time the current came up, it simply broke through even the second current limit, displaying no skipped pulses thereafter, continuing its steady rise into oblivion. I actually captured the entire sequence right till the moment the IC disintegrated (the scope plots went wild thereafter). Of course I knew by then that if the layout for this extremely fast-switching (Fet-based) switcher family was not good, sometimes customers would manage to hit the second level of current limit protection inadvertently (simply due to injected noise). In that case they would typically come back complaining of not being able to get full load from their ICs. And I would recommend the 0.1µF ceramic capacitor right at the input. But this was a blowup the likes of which I had never seen before.

Note that this is an entirely different problem from the $D > 50\%$ foldback issue discussed elsewhere in this book, because that is really related to a rather weirdly designed *first* level current limit of this family of ICs.

When I had initially looked at this Japanese customer's board, I already knew the layout was not very good because the freewheeling diode was a little too far away from the IC—maybe just three-quarters of a centimeter away, but enough to encounter the second level current limit (I knew that that was a possibility by then). But what was really surprising was that the second current limit worked only the *first* time around. Eventually, the problem, it turned out, was not just the PCB layout, but the fact that the customer was using the *through-hole version* of the device mounted on to an extrusion heatsink. So the leads of the device were already long enough to *almost* start causing trouble—because however hard you try to put them close to the IC, the input decoupling capacitor would still be two full lead lengths away, *and so would the output diode*. So, strictly speaking, the situation only got *worse* by putting the diode a little further away on the PCB. The device was marginal even to start with. But it was clearly enough to cause complete destruction. I thus realized that if I had a chance, I would never use the through-hole versions of this family for this reason alone. The SMD version has much shorter leads and you therefore start off with a great advantage. The only problem with the SMD package is its thermal dissipation capability (no heatsink is possible obviously, we have to rely on PCB copper planes).

Note that bad input decoupling was not the primary reason for the failure here, because that was known to only cause the first current limit to be breached on occasion. This breach of the second current limit was due to the ringing noise *at the SW node* gaining ingress into the IC, and causing a far more serious problem. No amount of input decoupling could have helped here.

My final recommendation to the customer was to put in a small RC snubber very close to the IC between its SW and GND pins. Typical values of this snubber are 470pF to 4.7nF and 10Ω to 100Ω. Note that since this Band-Aid fix is very layout and parasitic dependent, I usually ask the customer to try all the corner combinations first, such as 470pF/10Ω, 470pF/100Ω, and so on. The customer may also need to play with intermediate R and C values to optimize performance and not take too big a hit in efficiency in the process. I know the Japanese customer evaluated the snubber fix at his end and went into full production with it.

Boost PCBs

In Figure 6-2, we do similar trace section analysis for a Boost. We thus realize that only the output section needs to be looked at closely for this topology. Of course, as mentioned previously, we should not forget the needs of the control sections of the IC. So the input rail to that (not shown in Figure 6-2) needs its own decoupling (typically *RC*-based). For all topologies, that is in fact a key requirement—that the control IC be powered off a clean rail. It is just that in a Buck, the input decoupling capacitor for the power stage and the decoupling capacitor for the control are often the same component. Though in some cases, it may be necessary, even for a Buck controller IC, to add a small RC going to its supply pin (typically a 10Ω resistor and an additional 0.1μF ceramic capacitor).

We are also seeing another pattern emerge here, that the inductor and its associated traces are *not* critical in *any* topology. That is because the inductor smooths out the current through it, so obviously no "edges" of current pass through it. The slowly undulating inductor current has ripple, but not noise! We need *not* pay very close attention to it, except to keep it away from sensitive nodes, in particular the feedback trace.

Buck-Boost PCBs

This topology is the hardest to implement at the layout stage, simply because both the input and output sections see a pulsating current waveform. That is clearly reflected in the trace section analysis shown in Figure 6-3. It is important to minimize almost all the current loops, both on the input side and the output. They all see an edge of current and will therefore complain in the form of voltage spikes. Again, the only exception, in principle, is the inductor. Unfortunately, we can't afford to keep it far away, because doing so would force us to make the length of its adjoining traces long too, and that is not acceptable, because they are still critical in this typology.

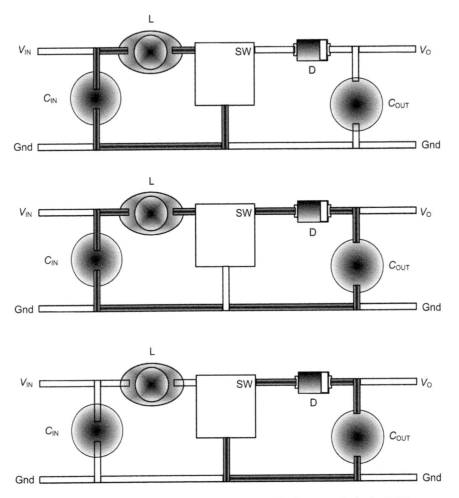

Figure 6-2 Boost from Top to Bottom—Switch ON, Switch OFF, Critical Traces

Note that in Figure 6-3, we have a negative to positive Buck-Boost. But in fact the very same arguments and conclusions apply to a positive to negative Buck-Boost. There is no difference in layout principles, except, of course, for the fact that "Ground" is different.

Forward Converter PCBs

Even though a Forward converter is a Buck-derived topology, things change somewhat in terms of PCB layout, by the inclusion of the transformer. A transformer operates on the principle of AC transfer, so it requires the currents on its Primary and Secondary windings to have sharp edges, or there would be no coupling at all! Note that its core may still "think" it is an inductor, because the field inside it remains smoothly undulating. But the

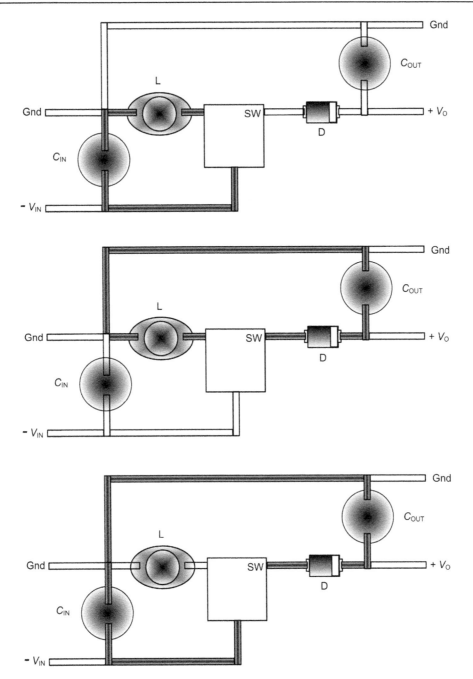

Figure 6-3 Buck-Boost from Top to Bottom—Switch ON, Switch OFF, Critical traces

windings tell a different story, because they chop the current waveform to make energy transfer possible. That is also why in inductors we simply use the DCR (DC resistance) to calculate the copper loss; but in transformers, we need to understand AC resistance effects to properly estimate its copper loss. However, core loss calculations are essentially the same for both an inductor and a transformer.

So in Figure 6-4, we do trace section analysis for a Forward converter, and find that there are *two* separate current loops we need to minimize here. The differences between Figure 6-4 and Figure 6-1 are subtle but important. The latter is in effect only one current loop, even though it spans both the input and output sections.

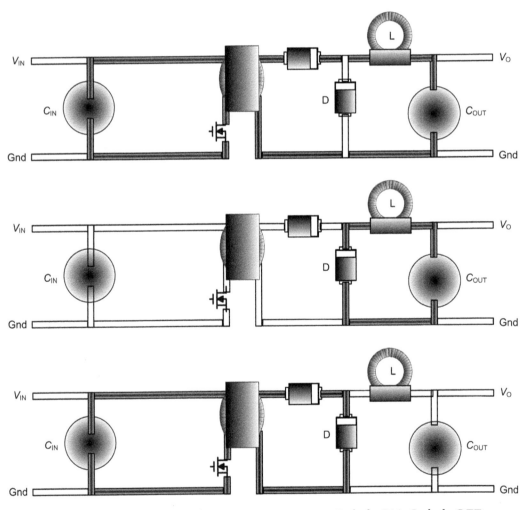

Figure 6-4 Forward Converter from Top to Bottom—Switch ON, Switch OFF, Critical Traces

Flyback PCBs

In Figure 6-5 we carry out analysis for a Flyback PCB, and realize that unlike a Forward Converter, even the output capacitor has to have very short interconnecting leads and trace lengths. That situation is similar to the Buck-Boost in Figure 6-3, from which the transformer-coupled Flyback is essentially derived.

The most stringent layout demands are thus made by the Flyback. It may be considered "cheap and dirty," but that epithet *cannot* be allowed to refer to its *PCB layout*! Otherwise its performance will certainly be as so described.

With everything required to be close to everything else, the question is, what is the priority list? The answer to that is *everything* is equally important, at least in a Flyback. However, remember that the secondary-side trace sections carry far more current than the primary

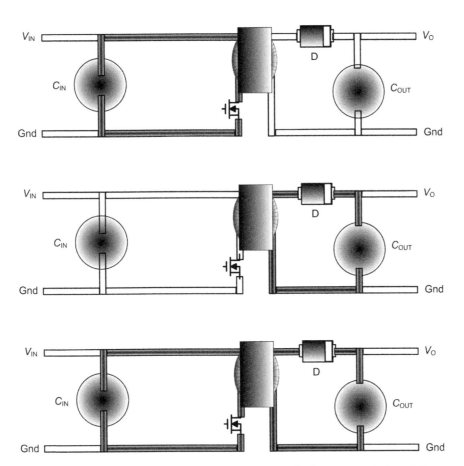

Figure 6-5 Flyback from Top to Bottom—Switch ON, Switch OFF, Critical Traces

sections. So if we switch 1A in 50ns versus 20A in 50ns, it is obvious where the highest dI/dt will occur! We clearly need to focus on the secondary-side trace sections the most. This is also true because any trace inductance here gets multiplied by the square of the turns ratio, and reflects into the primary side, as discussed previously. This greatly increases the dissipation in the primary-side RCD/zener clamp and severely degrades the converter efficiency. We have to really struggle to minimize secondary-side inductances, especially for low output voltage rails, that is, those with higher turns ratios.

Some Points to Keep in Mind during Layout

Let us summarize these for quick reference purposes:

- During a crossover transition the current flow in certain trace sections has to suddenly come to a *stop*, and in certain others it has to *start* equally suddenly (within 100ns or less typically, which is the switch transition time). These trace sections are identified as the "critical traces" in any switcher PCB layout. A very high dI/dt is created in them during every switch transition. Expectedly, these traces end up "complaining" vociferously in the form of small, but potent, voltage spikes across them. We realize that this is just the equation $V = L$dI/dt playing its part, with the L being the parasitic inductance of the PCB trace. The rule of thumb for the inductance presented by a trace is *20nH per inch of trace length*.

  Once generated, these noise spikes cannot only appear on the input/output rails (causing related performance issues), but also infiltrate the IC control sections, causing it to behave anomalously, and unpredictably. We could even end up briefly losing the usual current limiting function too, leading to disastrous consequences.

- Mosfets switch faster than BJTs (bipolar junction transistors). The transition times of a Mosfet can be about 10 to 50ns, as compared to a BJT's typical transition time of 100 to 150ns. But that also makes the spikes far more severe in the case of converters that use Mosfet switches, because of the much higher dI/dt they can generate in the critical trace sections of the PCB.

  Note: One inch of trace switching, say, 1A of instantaneous current in a transition time of 30ns, gives a spike of 0.7V. For 3A, and two inches of trace, the induced voltage tries to be 4V!

  Note: It is almost impossible to "see" the noise spikes. First of all, various parasitics help limit/absorb them somewhat (though they can still retain the capability to cause controller upset). Further, the moment we put in an oscilloscope probe, the 5 to 20pF of probe capacitance can also absorb the spikes sufficiently, and we would probably see nothing significant. In addition, probes pick up so much normal switching noise through the air anyway that often we are never even sure of what we may be seeing!

■ Integrated switcher ICs (sometimes simply called switchers) have the switch in the same package as the control. Though that makes for convenience and low parts count, such ICs are usually more sensitive to the noise spikes generated by the parasitic trace inductances. That is because the switching node of the power stage (its swinging node, i.e., the one connecting the diode, switch, and inductor) is a pin on the IC itself, so that pin conducts any unusual high-frequency noise at the switching node straight into the control sections, causing controller upset.

■ Note that while prototyping, it is a bad idea to insert a current probe (through a loop of wire), anywhere in a critical trace section. The current loop becomes an additional inductance that can increase the amplitude of the noise spikes dramatically. Therefore practically speaking, it can often become virtually impossible to measure the switch current or the diode current individually (especially in the case of switcher ICs). In such cases, only the inductor current waveform can really be measured properly. Sometimes we can place a small sense resistor instead of a current loop, because a good resistor will not create inductive kicks at least.

■ In the Boost and the Buck-Boost, we see that the output capacitor is in the critical path. So this capacitor should be close to the control IC, along with the diode. A paralleled ceramic capacitor can also help, provided it does not cause loop instability issues (especially in voltage mode control).

■ Note that in the Buck and the Buck-Boost, the input capacitor is included in the critical path. That implies we need very good *input decoupling* in these topologies (for the power section). So, besides the necessary bulk capacitor for the power stage (typically a tantalum or aluminum electrolytic of large capacitance), we should also place a small ceramic capacitor (about 0.1 to 1µF) directly between the *quiet* end of the switch (i.e., at the supply side) and the ground—and also *as close as possible to the switch*.

■ We should remember that the control circuitry usually needs good *local* decoupling of its own. And for that we need to provide a small ceramic capacitor *very close to the IC*. Clearly, especially when dealing with switchers, the decoupling ceramic for the power stage can often do double-duty as the decoupling capacitor of the control too (note that this applies to the Buck-Boost and the Buck only, since a power input decoupling capacitor is only required for them).

■ Sometimes, more effective control IC decoupling may be required, in which case we can use a small resistor (typically 10 to 22Ω) from the input (supply) rail, going to a (separate) ceramic capacitor placed directly across the input and ground pins of the IC. This constitutes a small *RC* filter on the IC supply rail.

■ Note that in all topologies, the inductor is not in the critical path. So we need not worry much about its layout, at least not from the point of view of noise. However, we have to be wary of the electromagnetic field the inductor creates, because that can impinge on nearby circuitry and sensitive traces, and cause similar (though usually not so acute) problems. So generally, it is a good idea to try and use shielded inductors for that reason, if cost permits. If not, it should be positioned a little further from the IC, in particular keeping clear of the feedback trace.

■ The position of the diode is critical in all topologies. It leads to the switching node and from there on, straight into the IC (especially when using switcher ICs, because the SW node is then a pin). However, in Buck converter layouts in which the diode has unfortunately been placed a little too far away from the IC, the situation can usually be rectified even at a later stage, by means of a small series *RC* snubber connected between the switching node and ground (across the catch diode, close to the IC). This RC typically consists of a resistor (low-inductive type preferred), of value 10 to 100Ω, and a capacitor (preferably ceramic), of value about 470pF to 4.7nF. Note that the dissipation in the resistor is *twice* $1/2 \times C \times V_{IN}^2 \times f$ (since $1/2 \times C \times V_{IN}^2 \times f$ is dissipated in the resistor when charging the capacitor, and $1/2 \times C \times V_{IN}^2 \times f$ when discharging it across itself). So not only should the wattage of the resistor be appropriate for the job, but the capacitance should not be increased indiscriminately.

■ A first approximation for the inductance of a conductor (wire) having length *l* and diameter *d* is

$$L = 2l \times \left(\ln \frac{4l}{d} - 0.75 \right) nH$$

where *l* and *d* are given in centimeters. Note that the equation for a PCB trace is not much different from that of a wire.

$$L = 2l \times \left(\ln \frac{2l}{w} + 0.5 + 0.2235 \frac{w}{l} \right) nH$$

where *w* is the width of the trace. Note that for PCB traces, the inductance barely depends on the thickness of the copper on the board.

The logarithmic relationship above indicates that if we halve the length of a PCB trace, we can make its inductance halve, too. But we have to increase its width almost 10 times to get its inductance to halve. In other words, simply making traces wide may not do much—we need to keep trace lengths short.

■ The inductance of a via (through-hole) is given by

$$L = \frac{h}{5}\left(l + \ln\frac{4h}{d}\right) \text{nH}$$

where h is the height of the via in millimeters (equal to the thickness of the board, commonly 1.4 to 1.6mm) and d is the diameter of the via in millimeters. Therefore a via of diameter 0.4mm on a 1.6mm thick board gives an inductance of 1.2nH. That may not sound too much, but has been known to cause problems in switcher ICs, especially those using Mosfets, for which an input ceramic decoupling capacitor for the IC becomes almost mandatory. Therefore, it is strongly advised that this capacitor be placed extremely close to where the pins of the IC actually contact the board, and further, there should be no intervening vias between this capacitor and the solder pads of the pins.

■ Increasing the width of certain traces can in fact become counterproductive. For example, for the (positive) Buck regulator, the trace from the switching node to the diode is "hot" (i.e., swinging). Any conductor with a varying voltage on it, irrespective of the current it may be carrying, becomes an E-field antenna if its dimensions are large enough. Therefore the area of the copper around the switching node needs to be reduced, not increased. That is why *we need to avoid the tendency of indiscriminate copper filling.* The only voltage node that really qualifies for copper filling is the ground node (or plane). All others, including the input supply rail, can start radiating significantly because of the high-frequency noise riding on them. By making large planes, we also increase the probability of that plane picking up noise from nearby traces and components by means of inductive and capacitive coupling.

■ The so-called 1-oz board in the USA is actually equivalent to 1.4 mils (35μm) copper thickness. Similarly the 2-oz board is twice that. For a moderate temperature rise (less than 30°C) and currents less than 5A, we can use a minimum *12mils width of copper per amp for 1-oz board, and at least 7mils width of copper per amp for a 2-oz board.* This rule of thumb is based on the DC resistance of the trace only. So to decrease its inductive impedance and AC resistance, higher trace widths may be required.

■ We have seen that the preferred method to reduce trace inductance is to reduce length, not increase width. Beyond a certain point, widening of traces does not reduce inductance significantly. Nor does it depend much on whether we use 1-oz or 2-oz boards. Nor if the trace is unmasked (to allow solder/copper to deposit and thereby increase effective conductor thickness). So, if for any reason, the trace length cannot be reduced further, another way to reduce inductance is by *paralleling the forward and return current traces*. Inductances exist because they represent stored magnetic energy. The energy resides in the magnetic field. Therefore conversely, if the magnetic field could be canceled, the inductance vanishes. By paralleling two current traces, each carrying currents of the same magnitude but in opposite direction, the magnetic field is greatly reduced. These two traces should be parallel and very close to each other on the same side of the PCB. If a double-sided PCB is being used, the best solution is to run the traces parallel (over each other) on *opposite* sides (or adjacent layers) of the PCB. These traces can, and should be, fairly wide to improve mutual coupling and thereby the field cancellation. Note that if a ground plane is used on one side, the return path automatically images the forward current trace, and produces the sought after field cancellation.

- In high-power offline Flybacks, the trace inductances on the secondary side reflect on to the primary side, and can greatly increase the effective primary-side leakage inductance and degrade the efficiency. The situation gets worse when we have to stack several output capacitors in parallel, just to handle the higher RMS currents. Long traces seem inevitable here. This has been discussed in detail previously.

- With multilayer boards, it is a common practice to almost completely fill one layer with ground (if so, it should preferably be the layer immediately below the power components). There are people who, usually rightly so, consider the ground plane a panacea for most problems. As we have seen, every signal has a return, and as its harmonics get higher, the return current, rather than trying to find the path of least DC resistance (straight line), tries to reduce the inductance by imaging itself directly under the signal path even though that may be zigzagging away on the board. So by leaving a large ground plane, we basically allow nature to do its thing, searching and finding the path of least impedance (lowest DC resistance or lowest inductive impedance, depending upon the frequency of the harmonic). The ground plane also helps thermal management as it couples some of the heat to the other side. The ground plane can also capacitively link to noisy traces above it, causing general reduction in noise/EMI. However, it can also end up radiating if caution is not exercised. One way this can happen is to have *too much* capacitive coupling from noisy traces. No ground plane is perfect, and when we inject noise into it, it may be affected, especially if the copper is too thin. Also, if the ground plane is partitioned in odd ways, either to create thermal islands, or to route other traces, the current flow patterns can become irregular. No longer can return paths in the ground plane pass directly under their forward traces. The ground plane can then end up behaving as a slot antenna too, in terms of EMI.

- The only important *signal* trace to consider is usually the feedback trace. If this trace picks up noise (capacitively or inductively), it can lead to slightly offset output voltages, and in extreme cases (though rare), even instability or device failure. We need to keep the feedback trace short *if possible* so as to minimize pickup and keep it away from noise or field sources (the switch, diode, and inductor). We should never pass this trace *under* the inductor, or *under* the switch or diode (even if on opposite sides of the PCB). We should also not let it run close to and parallel, for more than a few millimeters at most, to a noisy (critical) trace, even on adjoining layers of the board. Though if there is an intervening ground plane, that should provide enough shielding between layers.

  Keeping the feedback trace short may not always be physically feasible. We should realize that keeping it short is certainly not of the highest priority. In fact, we can often deliberately make it long, just so that we can assuredly route it away from potential noise sources. We can also judiciously cut into the quiet ground plane to pass this particular trace through, so that it is, in effect, surrounded by a "sea of tranquility."

Thermal Management Concerns

Larger and larger areas of copper do not help, especially with thinner copper. A point of diminishing returns is reached for a square copper area of size 1in. × 1in. Some improvement continues up to about 3in. (on either side), especially for 2-oz boards and better. But beyond that, external heatsinks are required. A reasonable practical value attainable for the thermal resistance (from the case of the power device to the ambient) is about 30°C/W. That means a 30°C rise for every Watt of dissipation inside the IC.

To calculate the required copper area, we can use as a good approximation the following empirical equation for the required copper area:

$$A = 985 \times R_{TH}^{-1.43} \times P^{-0.28} \text{ sq in}$$

Here P is in Watts and Rth is the desired thermal resistance in °C/W (degrees Centigrade per Watt).

For example, suppose the estimated dissipation is 1.5W. We want to ensure that, at a worst-case ambient of 55°C, the case of the part does not rise above 100°C (safe temperature for the PCB material—do not exceed!). Therefore the Rth we are looking for here, is

$$R_{TH} = \frac{\Delta T}{P} = \frac{100 - 55}{1.5} = 30°C/W$$

Therefore, the required copper area is

$$A = 985 \times 30^{-1.43} \times 1.5^{-0.28} \text{ sq in}$$

$$A = 6.79 \text{ sq in}$$

If this area is square in shape, the length of each side needs to be $6.79^{0.5} = 2.6$in. We can usually make this somewhat rectangular or odd-shaped too, so long as we preserve the total area. Note that if the area required exceeds 1 square inch, a 2-oz board should be used (as in this case). A 2-oz board reduces the thermal constriction around the power device and allows the large copper area to be more effectively used for natural convection.

We should not think that heat is lost only from the copper side. The usual laminate (board material) used for SMT (surface mount technology) applications is epoxy-glass FR4, which is a fairly good conductor of heat. So some of the heat from the side on which the device is mounted does get across to the other side, where it contacts the air and helps reduce the thermal resistance. Therefore, just putting a copper plane on the other side also helps, but only by about 10 to 20%. Note that this opposite copper plane need not even be electrically the same point; it could for example just be the usual ground plane. A much greater reduction of thermal resistance (by about 50 to 70%) can be produced if a cluster of small vias (thermal vias) are employed to conduct the heat from the component side to the opposite side of the PCB.

Thermal vias, if used, should be small (0.3 to 0.33mm barrel diameter), so that the hole is essentially filled up during the *plating* process. Too large a hole can cause solder wicking during the reflow soldering process, which leads to a lot of solder getting sucked into the holes and thereby creating bad solder joints for parts in the vicinity. The pitch (i.e., the distance between the centers) of several such thermal vias in a given area is typically 1 to 1.2mm. A grid of several such vias can be placed very close to, and alongside, a power device, and even under its tab (if present).

Making Boards Suitable for Troubleshooting

So do we make eval boards or demo boards? I still don't know the answer to that! But I can tell you some of the things I personally do that make for easy troubleshooting. The way I do it, by concealing some *invisible options*, most boards I build can go straight to the customer. He or she will never know the difference!

The first thing I like to do is to try and create a common ground island between the input and output. It is shaped more like a U (see Figure 6-6). I am still not very comfortable with *linearly* laid out designs. My other preferences are shown all together in Figure 6-6. In

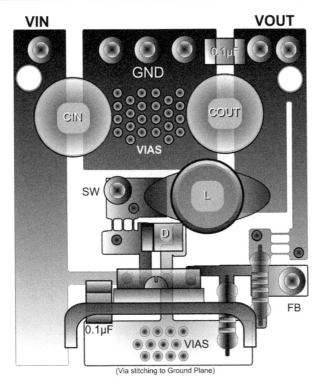

Figure 6-6 A Recommended Buck Switcher Layout on a Double-sided Board

subsequent figures, we will zoom in on various parts of this PCB and see the specific recommendation arising from that view.

Recommendation 1 (Figure 6-7): Several ground prongs are placed on the input/output ground island. If we have ground terminals all over the board, we can create a ground loop as shown in Figure 6-8. Of course we can choose to reference our scope probes at any point along the ground plane, but we must make sure that *all* the scope probe ground clips are at *one* single point at any given moment. I often solder an inch of solder wick at the ground location I prefer (usually right at the output, but often at the ground pin of the IC), and then I can keep clipping scope clips on to it. It doesn't look high tech, but it works.

Recommendation 2 (Figure 6-9): As previously discussed, a small ceramic capacitor of about 0.1μF is recommended at the output, for carrying out more meaningful noise and ripple measurements. That measurement is tricky and we must do our best to avoid any pickup by removing the ground lead of the probe. The scope probe must be used correctly as shown. Also disconnect all other scope probes completely from the board during this measurement.

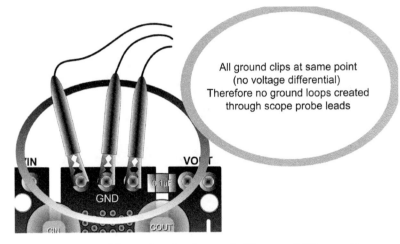

Figure 6-7 Avoid Putting in Ground Clips at Different Parts of the Board

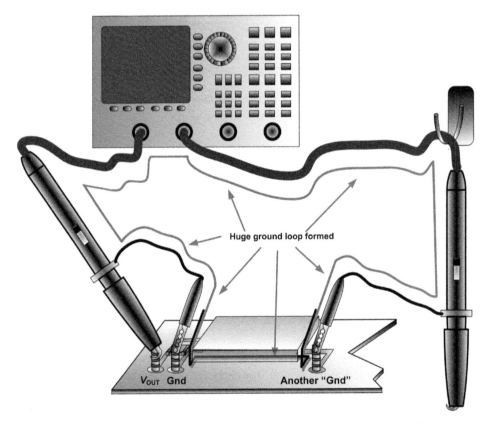

Figure 6-8 Ground Loops Created by Placing Ground Clips at Different Points Along the PCB

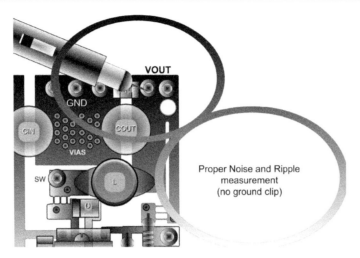

Figure 6-9 Putting a Small Ceramic Capacitor Between Closely Spaced Output Prongs Allows for a Proper Measurement of Noise and Ripple

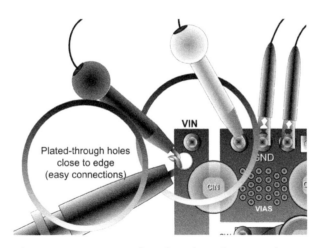

Figure 6-10 Leave Plated Holes Close to the Edge of the PCB for Quick Connections of Supply Clips and Probe Tips

Recommendation 3 (Figure 6-10): I like to leave plated through-holes along the edge of the board to allow for quick connections (using the clips from the bench power supply leads, or from the load). I prefer to avoid soldered prongs completely. But I also make the hole the right size, so if a prong is really needed, the hole will accommodate that too. All the eval boards I made for my companies never had any prongs on them. That not only saves a surprising amount of money, but it also gives the customer flexibility in choosing his or her preferred type of connection. Of course I would prefer the customer to actually take the trouble to solder the connections down firmly on the board. The holes come in handy for that, too.

Recommendation 4 (Figure 6-11): Of course the basics of a Buck switcher layout must be followed. The diode must be very close to the IC. Its cathode goes to a test prong labeled SW. We realize that almost everyone wants to look at this node, and it must be made available.

Recommendation 5 (Figure 6-12): The bulk capacitor has been placed close to where the input supply leads came in, but the ceramic decoupling capacitor must be very close to the IC as indicated here.

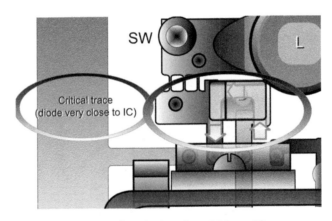

Figure 6-11 Catch Diode Placed Very Close to the IC, Between SW and GND Pins

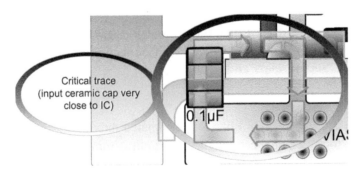

Figure 6-12 Input Decoupling Capacitor Placed Very Close to the IC Between VIN and GND Pins

Figure 6-13 Output Sensed Remotely for Better Regulation

Recommendation 6 (Figure 6-13): As discussed in Chapter 2, the output needs to be sensed remotely for best regulation.

Recommendation 7 (Figure 6-14): The divider needs to be physically close to the IC to avoid noise pickup along the feedback trace. The feedback trace doesn't run under the inductor or diode in particular, and is in fact kept at least a couple of millimeters away from the body of the diode. Also note that a test point has been created for this node, too.

Recommendation 8 (Figure 6-15): All the grounds are stitched generously together at various points as shown. Note that we should *not* try to connect the various grounds on the

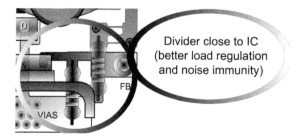

**Figure 6-14 Voltage Divider Placed
Physically Close to the IC to Avoid Noise
Pickup Along Feedback Trace**

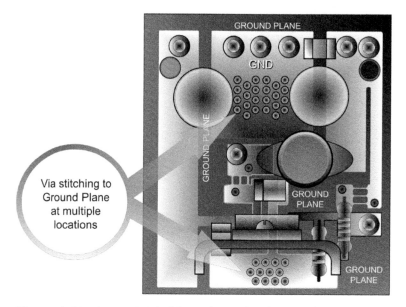

Figure 6-15 Ground Stitching Through Clusters of Vias

component side with traces (in addition to their connection via the ground plane), because that could *create ground loops*.

Recommendation 9 (Figure 6-16): The board is close to simultaneously meeting the requirements of a typical Apps engineer and a more investigative customer. But at this point a digression occurs (however unnoticeable it actually is). Many engineers want to leave a provision for inserting a current loop and carrying out a Bode plot, for example. But such evaluation options can make the board look very clumsy and unfit to send to the customer. So what I personally do is to simply leave three or four fine trace *interconnects* along with two small plated through-holes on either side of this mesh. When the time comes, I can

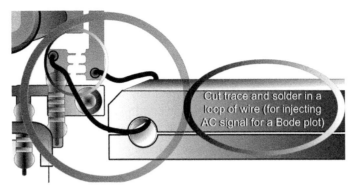

Figure 6-16 Leave Options on the Board for Easy Connection to the Loop Analyzer

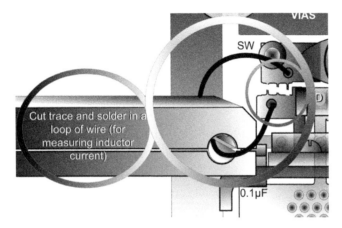

Figure 6-17 Leave Options on the Board for Easy Connection to Current Probe for Measuring Inductor Current

easily cut the interconnecting traces with a small X-Acto knife and solder in a wire loop connection through the holes as shown. Yes, we can use a passive probe for injecting the signal for a Bode plot—a transformer or sense resistor may never really be required. Just make sure the probe is really *passive* (i.e., an AC probe that goes directly into the scope, not via a current probe amplifier).

Recommendation 10 (Figure 6-17): Similar to Recommendation 9, I leave a provision for inserting a current probe to monitor the inductor current.

That is how most of my board designs can go straight to the customer, and also be used fully for a normal Apps evaluation. The perennial question "Is it an eval board or a demo board?" has been virtually sidestepped.

But there is one more recommendation that has not been illustrated, relevant to a four-layer board. My preference is to have all the components on the top layer and the layer right below it as the ground plane. For the remaining two layers, I would rather have the inner layer as the one for routing and general interconnections. That makes the outermost layer the second ground plane. This provides some shielding and helps make the board very "quiet" inside the customer's system. The two ground planes are of course to be stitched together at multiple points. However, I generally order *two* versions of the board—one with the customer-friendly (and EMI-friendly) arrangement of layers discussed above, and one where *both* the ground planes become internal layers. I do that because during evaluation, if I want to cut traces or perform some experiments, it always helps to have the *routing layer available outside*. If the routing layer is internal, there is little you can do to change anything on the board during troubleshooting.

Working without a Ground Plane

384x-based Controllers on Single-sided Boards

Suppose I take the greatest gift known to the world of switching power supplies, and ask you to build a high power AC-DC power supply *without it*? Wouldn't you feel really challenged? Because if you did, you wouldn't be alone! Many AC-DC commercial power supplies are still built on cheap single-sided CEM1 or CEM3 PCB laminate and therefore can't afford the luxury of the *magical ground plane*. Bereft of that dependable ally, routing requires specialized skills, developed through years of experience. If anything, it makes you keenly aware of all that the ground plane accomplishes for you in multilayer boards, without you even being aware of it.

Let us take the popular 384x family for the purpose of illustrating some key routing principles in AC-DC power supplies.

The first thing we should be conscious of in laying out our design is that such ICs always have a current sense resistor connected to the Source lead. So it is not possible to minimize the physical distance between the Source and the Mosfet ground. We will certainly use noninductive sense resistors and so on, but there will still be some bounce left. To avoid aggravating the situation further, we need to at least minimize the bounce between the IC ground and the Mosfet ground. We also have to remember that the Fet will usually be a certain distance away from the IC, on a heatsink, but we still want to minimize Gate drive trace impedances if possible.

So as in Figure 7-1, we should first create a small (local) ground island next to the IC. All the circuitry around the controller is referenced to this ground. For example, the timing capacitor, the IC input supply capacitor, the I_{SENSE} filter capacitor, and so on, are all connected here.

We also have a larger (power) ground island right next to the Fet. The negative terminal of the input bulk capacitor must also connect here. Its positive terminal is connected to the transformer primary winding. The other end of that winding goes straight to the Drain of the Fet. It is important to minimize the area of the loop carrying primary-side current, because otherwise it radiates severely. More on that in the next section.

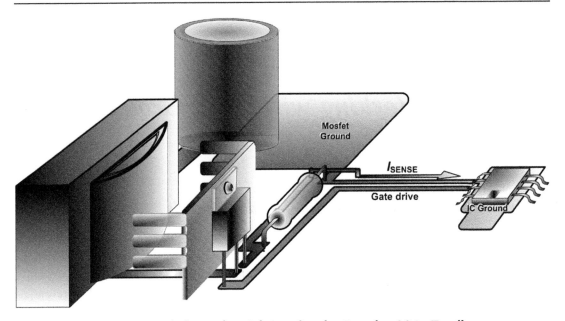

Figure 7-1 A Recommended Routing Scheme for the Popular 384x Family

The IC ground and Fet grounds are now connected through the fairly long trace shown in the figure (of course we would like to make this as short as possible, but sometimes we just can't). *We must ensure there are no other inadvertent connections between these ground islands; otherwise a ground loop will be created* (which could end up carrying circulating currents and thereby radiating).

To minimize spurious Gate drive signals from being applied, we run the Gate drive trace rather close and parallel to the connecting ground trace. So in effect we are trying to introduce mutual coupling between the forward and return traces. This will reduce the inductance in the Gate drive. But it will also reduce any bounce between the two ground islands, since the Gate driver in the IC is mainly responsible for all the spikes of currents flowing in the connecting ground trace. The IC has a few more connections to the outside world, namely

1. The I_{SENSE} signal picked up from the sense resistor. Note that this is a sensitive trace and under no condition should you run it alongside the Gate drive trace, as it will pick up noise and produce a fairly staggering amount of jitter. That is why in Figure 7-1, it has been jumpered and made to run alongside the much quieter ground trace.

2. The feedback signal. This usually comes from the opto-coupler. We do try to keep it away from noisy traces and components, but in reality, it is not that prone to noise pickup as many tend to instinctively believe.

In general, it is OK to have a quiet trace run past a noisy trace at a 90° angle. Running them parallel even for an inch or so can produce interaction and noise coupling. We should also try *not* to run sensitive traces *under* noisy or radiating components, even if we have to go a little further around it. In general, keep sensitive traces away from magnetic components, freewheeling diodes, and switches.

Watch the Primary Side Current Loop

Many years ago in Singapore, we were in the process of designing a new Flyback for a major computer manufacturer (known for its alternative operating systems). The metal box for the power supply, including all mounting locations, vents, holes, and so on, was predetermined. Somehow we had to fit within the constraints.

One of the problems we faced was a mounting standoff on the primary side that we just couldn't get around satisfactorily. We had EMI filters and other items taking up all remaining space. The only way out was to position the bulk capacitor and the switch on opposite sides of the standoff, as shown in Figure 7-2. Unfortunately, these long traces were

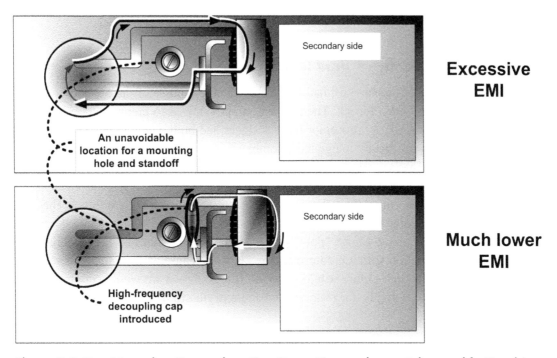

Figure 7-2 Box Mounting Constraints Can Force Non-optimum Primary-side Routing and Cause Excessive EMI—Possible Solution May Be to Mount a Decoupling Capacitor Close to the Fet

in the path of the sharply edged primary-side switch waveform. In effect we now had lousy input decoupling (for the power stage). Also notice the rather large current loop in the upper half of the figure. We really had a good antenna too! Bringing the EMI and output ripple into compliance on this power supply was the biggest struggle we had while I was there. We did it somehow, almost strangulating ourselves in the process. Now when I look back at this incident, I wonder why we didn't place a ceramic decoupling capacitor close to the switch, as shown in the lower half of the figure. The bulk capacitor could have successfully managed to provide the low-frequency current components, whereas the high-frequency capacitor could have really decreased the effective loop area in which the high-frequency components were circulating.

And the Secondary Side Current Loop, Too

In a Flyback, the high-frequency current loop encloses the transformer secondary, the output diode, and the output capacitor. This loop must be minimized as far as possible.

In a Forward converter, the high-frequency current loop encloses the transformer, the output diode, and the freewheeling diode. This loop must be minimized. Note that the output choke and output capacitor see relatively smooth (low-frequency) current, so their positioning is not critical.

Schottky Diode Failures—A Bead to the Rescue

In my Singapore-based company, we had some inexplicable failures of the 5V-output Schottky diode of our 70W Flyback. The product had just moved to preproduction, and the nervous owner-cum-CEO-cum-senior-most designer had built a couple of thousand power supplies in his "vertically integrated" Bombay factory for various life tests, burn-ins, and so on. That's when we got a call that about 2 or 3 units had failed mysteriously. Of course the switching Fets had also failed (that was always a given), but it was surprising that this time, the Schottky diodes were also found to have failed. The good thing about this company was that they recognized that even one failure in a thousand units constituted a disturbingly large "ppm" rate, one that they couldn't afford to put aside and go into full production with. Of course if they knew the cause and eventually declared it to be non-chargeable, that failure wouldn't count. But until then we would certainly be spending many sleepless nights, far away in Singapore, trying to troubleshoot this new failure mode.

I need to emphasize that this company didn't *ever* put failures like these aside, by blaming either their admittedly overworked and underpaid production staff, or the humidity, or a random bad part, or even the Gods. They were invariably known to send every failed part to their respective vendors for an immediate failure analysis report. Call them paranoid if you like, but they sure made a killing that way, with almost every account they held. And this was no exception.

Remember, only three units had failed so far (in Bombay), and the unit we had on hand in Singapore had obviously not failed (that's why we were able to study it!). Replicating the failure on the bench successfully *and* managing to find enough clues in the few μs leading up to the failure was nearly impossible. We could actually do neither!

So we had to first try to find some *correlation* here. The questions we asked, the rationale behind them, and their replies are detailed below:

a) Did it fail during power-up or shutdown? That is usually one of the most likely moments of failure of a Flyback, since a Buck-Boost topology always has the highest peak current at the lowest input voltage. So any overstressed part is most likely to succumb when the input is rising from zero to a higher steady value, or powering down. Usually we depend on the cycle-by-cycle current limit and/or the duty cycle limit and/or the undervoltage lockout to save the show. But the tolerances of these protection thresholds are not very accurate, and also drift over time. So that is a possibility. *Answer:* But no, not in this case.

b) Did it fail at high input voltages? Typically, most power components will see the highest voltage stress at high input voltages. *Answer:* No, not in this case.

c) Did it fail "cold" (room temperature) or after running for some time? High temperatures are responsible for increased failure rates over a long period of time. But a transistor or diode doesn't fail *instantly* the moment you cross 150°C. However, a Fet could go into rapid thermal runaway, since its Rds increases steeply with temperature. But we had an OTP thermistor glued on to the body of the Fet, and that was working properly (it also seemed intact on the failed units). *Answer:* No, not in this case.

d) Did it fail under abnormals? These are typically sudden applications of load current beyond the maximum rating (note that a proper output short circuit is usually the most benign overload case—the worst is at the load current just *before* the point where the output starts to fold back). *Answer:* No, not in this case.

Finally, there seemed to be *no pattern* to the failures! So we had to hook up a scope and current probe to check the 5V diode waveforms to rule out excess current and voltage stresses. Note that a freewheeling diode failure will always precede a Fet failure, very rarely the other way around. In other words, if the diode failed, we would expect the Fet to fail soon thereafter, but if the Fet was what started it all, the diode would usually be found intact. So at least we were reasonably sure we were heading in the right direction by looking at the diode, not the Fet! We looked at all the diode waveforms, and we were sure we still had enough derating margins. Under no condition were we even close to the ratings of the device as far as we could see. Then we asked another question:

e) We had two approved diode vendors. Were we getting failures with one vendor and not the other? *Answer:* Yes, it is true, the cheaper diode was the one failing. Aha!

So now we started sifting through their respective datasheets with a fine tooth comb, looking for differences. It is then we spotted a likely candidate. We suddenly realized that a Schottky diode has a dV/dt rating, too, in addition to its reverse voltage and forward current ratings. The cheaper diode had a published dV/dt rating of about 2000V/µs, whereas the high-quality (expensive) diode had a rating of 10,000V/µs. Big difference. So you can ask—did we just blindly spend 30 cents or so more and go with the expensive diode from that point on? No way. That would've come out of our pockets at this stage. And it was a substantial increase considering the entire power supply was being sold to the customer for about $15 to $20 (I can't remember exactly). So we *had* to find a way of using the existing diode with minimum changes (to either the design, which had already been "pre-qualed" (pre-qualified) by the customer, or the BOM cost). Tall order.

We hooked ourselves to the scope once again to look closely at the diode waveforms and see where we were on the dV/dt limits. At first sight we seemed well within the ratings. On closer examination, I noticed a tiny wiggle (see Figure 7-3). If I drew asymptotes, I could see a small region in that wiggle-zone where the instantaneous dV/dt was twice the rated value. It was being caused by "unavoidably poor" secondary side routing. For thermal reasons, the diode was chassis-mounted, and this had necessitated longer traces. The extra trace impedance, combined with a whole bunch of component parasitics, was causing the wiggle. But we were really in no hurry to identify these parasitics any further and/or to write an intimidating paper in IEEE about them. We just needed a fix, fast.

An interesting aside—at the same time we were looking at the diode waveform, so was the assigned qualification engineer at the customer end. He had a state-of-the-art lab tucked deep inside a massive production facility of his company, somewhere in Singapore too. So we were on the phone with him all day, sharing our findings and suspicions. Problem was he couldn't see the wiggle at all. He tried for almost two days, bringing in his colleague to help him, too, but there was no wiggle at his end.

We then realized there was a difference in the instrumentation we were using. Since we used to place a lot of reliance on analog scopes for most of our bench work, we were still using a scope of the famous TEK 2400 series. I remember that only when it was time to record waveforms for some bland customer report would we wheel in the TDS420 (that's still my favorite digital scope, however outdated!). Analog scopes don't lie. So at our end, we could see the wiggle very clearly every time. But the customer was using a very advanced Tektronix digital storage oscilloscope. And that was his problem. Only after a couple of days of fiddling with the knobs and menus could he coax the wiggle out of the

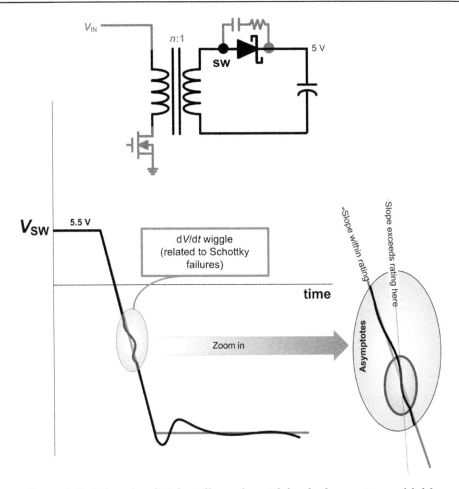

Figure 7-3 Schottky d*V*/d*t* Failures in a Flyback due to Unavoidably Poor Secondary Side Layout

waveform. And that was when daylight finally dawned in Singapore (we actually did work that late on days like this).

What was the fix? We finally ended up solving the problem by inserting a very small lossy ferrite bead into the secondary side high-frequency current loop. See Figure 7-4. We couldn't do away with the ringing (since it was related to some unidentifiable parasitics), but we could certainly damp it out. Note that such tiny beads are usually made of lossy (Ni-Zn) material, and their equivalent electrical model shows no measurable inductance, just a series AC resistance that peaks at very high frequencies. Remember, this bead was inserted in the *main* freewheeling path of the current. If we did anything more severe to impede the incoming edge of the magnetization current, we could create very serious

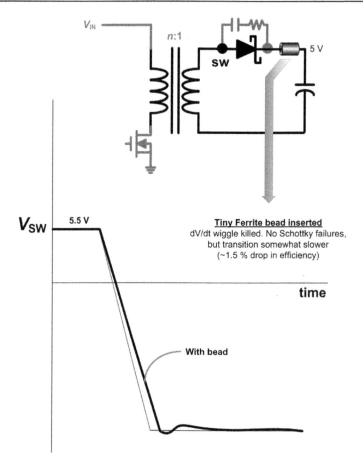

Figure 7-4 Damping Out the High dV/dt Wiggle by Inserting a Small Lossy Ni-Zn Ferrite Bead

problems indeed (Hint: never try to interrupt an inductor's stored energy!). All this bead did was to damp out the wiggle and slightly slow down the switching transition. The resulting dip in efficiency was about 1.5%, most of it as additional crossover loss in the Fet. But we rechecked, and all temperatures were still within our established derating margins, so we moved on. The bead must have cost a fraction of a cent. But all Schottky failures stopped completely once we implemented this tiny fix in production.

So what was really causing the wiggle? It was most likely non-optimum secondary-side layout. A Flyback is really trickier to lay out than, say, a Forward converter. But as you can see in Figure 7-5, this was possibly unavoidable in our case. Incidentally, this is the 5V/20A output. The diode dissipation can therefore be several Watts at high input voltages. Let's do the math! At 270VAC, the rectified DC is $270 \times \sqrt{2}$, which is about 400V. With a 20:1 typical turns ratio, we get 20V coming into the equivalent secondary-side Buck-Boost stage

Spreader plate needed to be attached
to box here, for thermal reasons

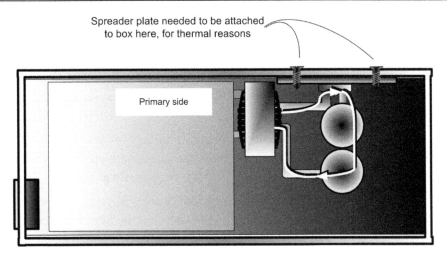

Primary side

Figure 7-5 Thermal Management Constraints Can Force Non-optimum Secondary-side Routing and Cause Excessive Ringing and Even Diode Failures—Possible Solution May Be to Put a Small Bead in Series with the Diode, to Damp out the Ringing

(read my *Design and Optimization* book for an explanation of that). From that point we have in effect a 20V to 5V conversion, which gives us a duty cycle of $V_O/(V_{IN} + V_O) = 5/25 = 0.2$. Therefore, the diode is conducting for 80% of the cycle (note that we have assumed continuous conduction mode here, which may not be the case). The conduction loss in the diode is therefore $0.5 \times 20 \times 0.8 = 8W$, assuming a diode forward drop of 0.5V. The only way to handle this heat, considering there was almost negligible airflow inside this power supply, was to attach the TO-220 (or DO-220) diode pack to an aluminum spreader plate and then screw that tightly to the chassis. But to guarantee good thermal contact, we had to put two screws *on either side* of the TO-220 device, as shown in Figure 7-5. All these measures necessarily pushed the diode further away from the transformer, thus necessitating rather long traces for the sharp-edged secondary current waveform. We were clear that this particular layout was almost certainly responsible for the wiggle we saw, because we had never seen it earlier, and later, even if we swapped components from another power supply model with better routing (and no observed wiggle), we couldn't get the problem to go away on this particular model. It was truly stuck to the specific layout. And that could not be changed.

The Real "Switch"

Of course there is no copyright or patent issue involved in showing you the fairly clever routing cum thermal management scheme of Figure 7-5. At least not today. But try to understand how delicate that situation was at that time (over a dozen years ago). My

Singapore-based company, along with a well-known $500 million revenue power supply company, were the only two companies picked as suppliers by our major OEM customer. We therefore competed ferociously with each other. The OEM would dictate the exact box design, mounting holes, and so on, beforehand, the only difference being on the inside. But that's where the margins were, and also our ability to grab larger volumes, as time went on, by reducing our prices further.

During the development stages, we needed to go several times for testing into the OEM's sophisticated qualification labs, tucked deep within their awesome production facility in Singapore. And that's where the real "switch" used to happen. Read on!

One day, my Boss and I entered the OEM security checkpoint carrying two power supplies (both obviously ours) for testing. We signed in and went into the labs. I had been told that the engineer assigned by the OEM to qual and ensure compliance with the spec was actually a very close "friend" of ours (he had been "worked upon" by the "I don't take no for an answer" owner of our company several times by now). Agreed, he remained fastidious about the job assigned to him, but was still friend enough to allow what happened next! That day, he conveniently left the lab at just the right time. And on cue, my Boss winked and switched our power supplies with the competitor's units lying rather conveniently on the shelf (to my total surprise). There was no way to tell the difference from the outside, so we had no trouble at the checkpoint on the way out. As far as the Malaysian security guard on duty was concerned, we still had our very own two-power supplies in hand. He just commented how many Indians he sees nowadays coming in from power supply companies, and waved us off with a smile. We came back to our lab and opened the competitor's units to see exactly where they stood in the development process. My Boss had apparently been doing this "switch" for years by then, and expanded further: "I agree these guys are very good with their Forward converters, but not with Flybacks. Their early designs sucked big time. But now over several revisions, they too have learned to mount their Schottky diode on the chassis just as we did months ago. Now their design has finally started looking very similar to ours." The irony of all this was, the competitor seemed to be keeping as good a watch on us as we on them. Maybe, just maybe, that friendly validation engineer was actually "friends" with the whole wide world! We would never know. Though he did get laid off soon after. Last heard he was enjoying his tidy severance package at a sunny Singaporean golf course.

I can't seem to remember if wiggle failures were ever seen on the competitor's unit. Maybe not, because we were the ones really cheap (and *clever*) about it—we *would* knowingly use Schottky's with low dV/dt ratings (albeit with a bead), and similarly we *would* use only a 600V Fet (with very careful transformer design and input feedforward), while maintaining remarkable reliability and quality (as acclaimed in writing by the OEM on several occasions). I believe our huge competitor never really became clever enough with Flyback designs to do away with their expensive 800V Fet, for example. In fact, even the high-voltage monolithic switchers from various companies of today haven't dared to go below a 700V Fet. My Boss also mentioned that most Japanese power supply companies were (at that time) still using cushy 1000V Fets for the same universal-input application! I think I did learn a lot in Singapore—about really *switching* power supplies (deep inside validation labs), besides some other stuff too!

Home-Grown Strategies in Troubleshooting

Peeling the Onion

As you can imagine, there is no way I can write about every conceivable problem that may arise. And for sure, I haven't seen all the possible problems either. That's what makes power so much fun. But maybe I can share a tip or two with you. Because even though we all learn a little more every day, all we are really destined to see eventually is a *tip*—the tip of the iceberg. Luckily, we do eventually develop a certain type of *logic* that can help us get through the next problem a little more easily. And that's what I am going to talk more about here.

Troubleshooting, to me, is akin to peeling an onion. And often thereafter, *reverse-peeling* it, reassembling it back from its peels, almost like playing a movie backwards. In simpler systems, like DC-DC converters, looking for clues during the peeling phase is practicable, but in more complex systems like AC-DC supplies, reverse-peeling is my personal preference. For that, I first take out everything superfluous very carefully, until I reach the very core of the circuit—the switching engine of the converter. I make sure that that is ticking away just fine. I might even need to peel a few steps deeper into that engine, by disconnecting secondary outputs, for example. Then I start systematically putting back everything else around it until the problem reappears. But be warned—there is some carefully considered trace-cutting and also some ill-considered cussing involved in this process. Also do not forget to retain the basic functionality in the process. For example, if you "lift" the diode going to the 12V output, you may need to hook up a bench power supply to provide this rail externally, so you continue to provide the necessary current for the 5V regulation opto-coupler to do its job. Remember: you shouldn't try to use the primary regulated rail to provide the current to the opto-coupler as well. That can lead to a weird loop response. You should have a separately regulated rail if possible, for the opto.

What do I mean by taking out *everything* superfluous? That could mean any external *circuitry* not directly linked to the core functionality (such as current limits, OVPs, crowbars, OTPs, etc.). But it can mean much more. For example, a few days ago I walked into the lab to talk to a junior colleague of mine. He happened to be looking at some minor issue on a small DC-DC converter board in front of him. It was meant for a Li-ion cell input, and set for 1V output. Suddenly, he started looking really puzzled. "Why is the input supply showing 2mA at *no load*? It should be only a few µA in this condition. Is the part suddenly leaky?" So to be doubly sure I physically disconnected the leads going to the electronic load, knowing by now that electronic loads can malfunction and that even at a setting of 0A they can draw up to a couple of mA (especially the high-power rated ones, and more so after a year or two in the field, or a month on my bench). But the input supply reading stayed rock steady at 2mA. Clearly, the electronic load was good, but we weren't. In my experience, at such moments we need to pull back and start thinking, rather than just plunging headlong on. We need to start applying some very basic logic. *Asking questions.* One of the first things we should always ask is, what can I possibly do to make the problem get better or worse? That usually gives a clue to the cause of the problem. For example, we can ask, how does the problem respond if we change the input voltage? Or what if we change the bench power supply altogether? Does the problem change with temperature? (For the latter, it is often convenient to use a hot air gun, or hairdryer, along with a HFC-134a ("Freon") canister; though you should be forewarned: that might unmask another problem that you haven't seen so far!). What if I change the diode, does the problem go away, and so on. We need to think of everything, *however unlikely it seems to be.* The clue we got in our above case was that the 2mA reading stayed very steady as we varied the input. If it were simply a quiescent current issue caused by some leaky structure between the VIN pin and Ground, this reading was likely to go up as we increased the output. But it didn't! And what was getting more suspicious to me was the fact that it was almost exactly 2.00mA. How come? At the corner of my mind I realized that *there is only one thing in the converter that also stays rock solid as the input increases.* You guessed it! It's the output. So in some mysterious way, could this be related to the output voltage or output rail? But we had already disconnected the load. I checked again. Nothing seemed unusual with the setup. We had a few innocuous-looking scope probes hooked on to the board, but that was all! Or was it? At that moment I started peeling the onion. I started blindly removing anything extraneous. So off came the scope probes one by one. And suddenly the problem went away! I had in the last step just removed the probe hooked to the output rail. The problem was in fact quite simple—the scope channel had been set up for a 50Ω input termination. And since the signal was being DC-coupled to it, we were getting 1V/50 = 2mA. The entire process above actually took less than a minute or two to diagnose and fix. You may ask, why had the engineer been using a 50Ω termination setting anyway? Because he had been trying to characterize the noise on the output rail, and for that, a 50Ω termination (AC coupled) is in fact recommended to avoid scope cable reflections from distorting the observed signal.

Asking the Right Questions

At every stage, we must learn to ask plenty of questions. Because if we don't have answers to all of them, or haven't even bothered to asked, how would we ever know they were the *right* ones anyway? Here is a likely list (see Figure 8-1).

1) Is what I am seeing truly abnormal?

2) Is my equipment misbehaving?

3) Is the reading/data real?

4) Is the supply rail to the IC properly decoupled?

5) Is the IC at fault by design?

6) Is a particular part defective?

7) Is a particular part wrong?

8) Is there some strange interaction with the load?

9) Is there some strange interaction with nearby circuitry?

10) Am I trying to achieve the impossible?

11) Am I doing something known to be risky (to start with)?

12) Am I somehow managing to inflict damage to the IC?

Figure 8-1 Twelve Questions You Must Ask

Question 1: *Is what I think I am seeing truly abnormal? Is it really a problem, or isn't it?* For example, if you are seeing an overshoot of 50mV on startup, there is no reason to worry about that usually, because that amount of overshoot is considered quite normal. In fact for a 12V output setting, that may even be considered remarkable. In other words, look at it in *percentages*, such as 50mV on top of 12V is an overshoot of 0.4%. On a 1.8V output it would be about 3%, getting to the point where you might get slightly worried, depending on the application of course. Same for the DC regulation level of the output. However, if the system board connected to the output of the converter is failing, you can be quite sure your output *did* overvoltage for some reason, however temporary that event may have been. In

163

that case, the problem is actually worse than you think. Now you have to try and capture the event first, before you try to understand its cause.

Yes, sometimes you may find the customer expecting what my colleague used to call "unobtanium." The customer may simply post a line in the product specification without fully realizing what it entails, or what it might cost. Negotiating politely with the customer is usually recommended if you are in doubt about the validity of a spec (unless of course the customer "hates wasting time with stupid questions," and all questions are, by definition, stupid). So if the customer is expecting 20 years life on his output aluminum electrolytics at a room ambient of 45°C, tell him that that will cost him. A good capacitor will usually have a life of around 2000 hours (83 days, operating 24 hours a day). This life number applies when the capacitor is passing its rated ripple current at an ambient of 105°C. So with the doubling rule every 10°C fall in temperature, at best we will get to 2000×2^6 which is 128 thousand hours. Five years is about 44 thousand hours, so even if we pass no current through the capacitor at all, we can't get up to 20 years of life, not under these ambient conditions. We might go looking for "5000 hour capacitors," and then apply enough derating on the ripple current (more capacitors in parallel) to get there. But the question is—is it really necessary?

Many years ago, when designing AC-DC Flybacks for a well-known company making computers with alternative operating systems, we were pleasantly surprised that all they wanted in their spec was an estimated life of 15 thousand hours (about 2 years) for their output aluminum electrolytics, versus 5 years for almost every other customer out there. But their engineer explained it to us quite succinctly. "Why should we ask for 5 years life? The customer is not likely to operate the computer for more than 8 hours a day. So our "2 years" would amount to 6 calendar years for him. In any case, we think that is more than enough considering that customers will typically change their computers once every couple of years."

But you could find customers at the other extreme of the spectrum too. In that case you may need to *tell them* honestly they do need to pay *more* attention to a certain problem you have spotted, and may in fact need to pay more to have it resolved. You can also find other vendors out there, knowingly *playing down* known problems to their customers, for whatever reason. In other words, you *always* have to be careful, whichever side of the table you are on.

Once, while working as an Apps engineer at this maker of high-voltage Flyback monolithic switcher ICs, I happened to express concern about the way they were selecting the input bulk capacitors for their evaluation boards. To this day, in my opinion, they are making several misleading/erroneous recommendations to their customers via their eval boards.

a) They recommend an input capacitor selection based on the magic figure of 3μF/W. Though that does, in principle, give the converter a holdup time of 20ms as is usually required, it should be clear that what you really need is 3μF per *input* Watt, not *output* Watts. The front end (where this capacitor really is) cares only about the power *it sees*, which is the *input* power. The relationship between input power and output power is through efficiency, which could be all over the place. So you should not simply pick 180μF for a 60W universal-input Flyback. At 70% efficiency, the input power is about 85W. So you need to pick 255μF.

b) The second mistake, in my opinion, was that a typical electrolytic may have an initial tolerance of ±20%. And in addition, a fall of 20% capacitance occurs by its end of life. Assuming you still want the equipment to meet 20ms holdup time (and a respectable life), you really need to start with a capacitor of nominal value about 40% higher than what your 3μF/W rule tells you. In other words, for the 60W Flyback, you may need to pick a capacitor of nominal value 255 × 1.4 = 357μF. You may get by with 330μF, but certainly not with 180μF! And I don't think that amounts to overdesign. Just good engineering. In fact we were explicitly asked to do so by one of our major customers, a well-known computer manufacturer headquartered in Cupertino, California, when I was working in Singapore.

c) This high-voltage IC company also did no calculations whatsoever to verify that the ripple current passing through the capacitors was indeed within their respective ratings. Chemicon, for example, warns you that the usual life prediction formula applies only if you don't exceed the rated ripple current.

d) Further, in their apparent hurry to present cute, nifty eval boards to customers (to propel the sales of their ICs), the company would also instruct its CAD person to put the components in "as close as possible . . . period." They also ended up rewarding him or her with small bonuses for that effort. But I clearly remembered in Singapore, when we looked at a very nicely performing commercial power supply made by world leaders "Delta," the first thing that caught our attention was how carefully they had tried to keep heatsinks and hot components physically apart from the electrolytic bulk capacitors (to avoid diminishing their life severely). They were right after all. We had learned a lot from them. So I went back to the inventor-VP (now CEO) and stated my opinion on the capacitor issue. His answer simply was "if it fails they just throw it away, who cares?" An early end to something you may have paid good money for? I felt a wee bit behind the times.

Question 2: Is my equipment somehow responsible? That is sadly often the case. For example, some electronic loads can show weird glitches in the load profile they present to the converter under dynamic conditions. For example, if we are doing step load testing from 10mA to 200mA, all may be fine. But if we go from 0mA to 200mA, and see an output overshoot/undershoot, it could also be because of the electronic load. We may need to do that test either with another electronic load, or even a resistive load (an actual resistor). We could also put in a current probe in the leads going to the electronic load and ensure that the electronic load is indeed going smoothly between the two set load levels. Never assume anything.

Once we were doing repetitive output short circuit tests on our small DC-DC board, and sure enough, we managed to destroy the switcher IC. But the IC had pretty good current limiting (we had already ascertained that). It turned out that the DC power supply had caused the blow up because its output had started careening around all over the place whenever we shorted (and un-shorted) the outputs of our converter board. It got worse when we did that at a *certain rate*. We then really realized that even a bench power supply has an internal feedback loop and can go into oscillations. We had thought only switchers were suspect! It seems that whenever we shorted the output, we dragged down the input rail (the output of the bench power supply). So the bench power supply kicked in aggressively (rather too aggressively), trying to correct the situation. It ended up overshooting and then undershooting repeatedly, until finally it exceeded the voltage rating of the switcher IC. We actually did have some clues that led us on the trail of this hitherto unsuspected culprit. First, the likelihood of damage decreased if we reduced the input voltage. Therefore, it seemed to be not a *current* overstress, but a *voltage* overstress. Second, we noticed the wobble on the input rail and tried to pass on that information better to the bench power supply, hoping it would correct it. This we did by connecting remote sense leads from the bench power supply to the input prongs of our converter. But that only made matters worse, as the bench power supply now saw even bigger swings at its output, and became even more aggressive (and lousier too). Its loop was apparently poorly designed (we couldn't get our *switchers* to behave that badly). That's when we realized that this was *not* normal behavior for a bench power supply—up until then we were still under the impression that *we* were somehow instigating this behavior. We quickly replaced the lab supply with an HP/Agilent one, and that showed no further problems at all.

Incidentally, whenever you undertake corrective action and things get worse, you are actually still very close to identifying the cause. Also, we should remember that certain problems depend on *timing*. So in our case above, the *rate* at which we applied the load transients was important.

In another case involving timing, my colleague was sitting and toggling the enable pin of our latest HV (high-voltage) switcher IC and discovered that if the disable command comes within a certain 100ns window of the switching cycle, not only does the switcher ignore that command completely, but loses regulation, too, for a few cycles. So the output would overshoot every now and then. The only way to catch this type of occurrence is to set the scope to trigger on the rising edge of the output voltage, at a level about 10 to 25% higher than its steady value (past its normal noise and ripple platform), and in single acquisition mode. Then keep doing everything possible to get it to trigger.

Similarly if you toggle the enable pin at a certain rate in many ICs you can get to see *current* overshoots too. This often depends on allowing the input capacitor to discharge below the UVLO threshold, but not to the point where it hits the internal POR (power on reset) threshold. Because in that case, if you suddenly enable the IC, it has no soft-start anymore, and you will hit max duty cycle, and possibly staircase if the current limit circuitry is not well designed.

Question 3: *Is the reading real?* Artifacts of measurement are indeed common. Sometimes my colleagues have asked me about a high-frequency noise spike on the output they were particularly concerned with. All I did was to remove the tip of the probe from the output and touch it exactly where its own ground lead clip was attached. If they had expected to see a nice quiet straight line at 0V, they were wrong. Because the same concerning spike was still visible, which simply meant it was just noise picked up by the probe itself. It wasn't real. So either we do a proper measurement of noise, or at least we need to mentally subtract this pickup from whatever we are seeing on the screen. This is not unlike a conducted emissions test where we first take a scan without the converter being powered up, just so we can be sure which spike is real and which is actually just a cocky shock-jock on some nearby FM station.

In another case, my colleague discovered this for himself—if you have several probes connected to a board, you should try as hard as possible to have all their ground clips at exactly the same ground prong of the PCB. Because if you don't, imbalances in the PCB ground can create sizeable circulating currents in the grounds of the probes themselves (ground loops), leading to truly amazing artifacts on the output noise and ripple, and sometimes even chaotic behavior on the part of the controller.

All these stories inspired an old colleague of mine to present a seminar to the company's FAEs (Field Applications Engineers) tentatively titled "Be Sure What You Are Seeing Is Really True?" (Admittedly, I was then asked by my supervisor to "get some of the Chinese out of it!"). We have to realize that the measuring instruments become part of our larger system whenever we hook them up to take some data. Therefore, it is wise to question not just the quality of the instruments themselves, but their *natural* interaction with the device under test. For example, the few picoFarads of probe tip capacitance may be enough to either quench oscillations or create a new one altogether, especially if we touch it to the *high-impedance* feedback pin (not the type of feedback pin in fixed voltage option switchers). In fact I have never managed to put a probe tip on this node for too long. Things just seem to happen when you do! Though, by putting a DMM across it, we can have more success, but only in getting to know the DC voltage on this pin. Portable instruments, incidentally, fare better in some cases, since they aren't connected into the ground wiring of the building (which eventually loops around and comes straight back into the ground clip of your scope or your bench multimeter).

Question 4: *Is the supply rail to the IC well-decoupled?* We have dedicated an entire chapter to this (Chapter 2). You must read that carefully. In brief, we must always ensure that the supply rail is clean enough before we give any credence to malfunction.

Question 5: *Is the controller/switcher IC at fault (by design)?* This is actually a fairly common occurrence. No semiconductor product is released without its fair share of shortcomings. These are usually known to the company at the time of release, with the internal understanding that there were some lessons learned, and these will be resolved when the time comes for the next Rev, or the next product. Fair enough! But there are several variations to this theme, some that you may need to be aware of as you seek answers to a particularly stubborn problem. The three main variations are that the company knows about the problem; the company knows about the problem but does not want to admit it; or the company does not know about the problem.

a) The company knows about the problem.

1. The company knows about the problem and mentions it in the electrical characteristics (EC) tables of the datasheet (though remember that a TYP, or typical value, is not guaranteed; only the MIN/MAX values are). This then becomes a guaranteed spec since experts have opined that only the electrical characteristics (EC) tables are truly part of the contract with the customer. The rest, it is argued, is just general guidance (especially the first page of any datasheet which is best described as hyperbole). This company seems very upfront and will likely give you all the guidance you may ever need.

2. The company knows about it and mentions it in the *first few pages* of the datasheet. This is an instance where the company is quite forthright about the problem, but perhaps not definite enough about its spread or its impact, to guarantee it in the EC table (or unable to test it). Fair enough. You can work with them if you suspect a widening problem.

3. The company knows about it and mentions it in some *remote* part of the datasheet. You should know that every company has by now keenly realized that most customers barely go past the first couple of pages of a datasheet anyway (it is mentioned in internal meetings *all the time*). That's when you should be wondering if the company is just trying to create some sort of liability alibi, and no more. You are not likely to get any more detailed information from them about the problem either, since that portion of the datasheet probably wasn't directed at you in the first place (it was meant for the courts). At this point you should consider if you are better off looking decisively in another direction altogether. Learn to recognize the signs.

b) The company knows about the problem but doesn't admit it.

1. Why should they? There is no guarantee that a part must switch without excessive jitter, for example. These are only implied expectations you may have when you buy an IC. Further, what exactly constitutes "excessive?" Every switcher has *some* jitter!

   There are also implied expectations such as if you short the output and release it, the part must remain undamaged. But check whether the datasheet even mentions short-circuit protection. If not, it may be implied (to you) but not guaranteed (by them). In fact I recently learned that a giant semiconductor company had regularly been supplying giant PMICs (monolithic and complex power management ICs) to a very well-respected Japanese manufacturer for their state-of-the-art digital cameras, but *none* of the outputs of the integrated DC-DC converters had any active current limiting whatsoever! They said the loads on such custom ICs are so well-known that they don't think they need short-circuit protection. They were apparently depending on some divine intervention in the form of parasitics and duty cycle combinations to save the show. But what about normal startup stresses, component failures, and the like? It was even more galling that the semiconductor company, in an unbridled effort to generate more revenue, simply rechristened the part, and released it as a standard product for anyone out there to buy. No particular changes were made in the datasheet, and not the slightest mention that current limiting was *not* present on this PMIC. It was your risk if you had assumed anything that wasn't stated.

   Similarly, you may think a switcher is *not* supposed to have any output glitches or overshoots during startup. But to some extent all switchers do. And so, what you consider unacceptable may no longer bear any relationship to what the company says is acceptable, considering they have now realized they may be looking at a potential recall of a few million units!

   Many years ago my company asked me to fix a problem with a whole bunch of 3844 ICs purchased from a specific vendor that weren't working properly. The "same" part from another vendor worked flawlessly. I have described the entire episode in the Appendix of *Switching Power Supplies A to Z* if you are interested. I did learn several lessons from this. *First,* seemingly similar parts from different vendors can behave very differently. The primary reason for this is not just different levels of design expertise existing in different companies, but different fabrication processes. Every process has its own quirks, strange behaviors, nuances, leakages, noise pickup and sensitivities, feedthrough and crosstalk, and on and on. So the part may carry the same number "3844" but it could be an entirely different animal. *Second,* I truly learned about *implied expectations.* The end result was that the 3844 part had excessive jitter, which manifested itself as an unacceptable increase in the output voltage ripple. But where in the datasheet did they ever promise: "Jitter < 10%?" It was an inferred expectation, but only *our* inference. Obviously the vendor disagreed when confronted. *Third,* I learned that with some creative bench/design skills you *can* manage to put a poorly performing part (a cheaper one!) to good use. I would try to do that to save some money, assuming the part otherwise performs quite well, and the vendor is not *consistently* dishonest.

c) The company really doesn't know about the problem.

1. Maybe because this is a newly released part. But the vendor and the customer are now obviously handcuffed tightly together as they go up a steep learning curve. For all you know, looking back, this may have been the very part that ultimately caused the company's soaring stock to nosedive on the Nasdaq. Hopefully, you weren't on board at that time. In particular, you should stop dead in your tracks if the company tells you this is a new *process.* Because all their previous experience, device models, and so on, were based on their previous process. Now you should expect the unexpected. You might see ESD structures fail mysteriously, outputs go suddenly out of regulation after a few months of operation (e.g., the zener drift/mismatch issue that plagued the third-gen 267x Buck family in its early Revs), and so on. I personally would always prefer a mature part, even if its performance is not considered state-of-the-art or "best in class." Call me a little distrustful of all the marketing hype if you wish. But I learned that, for example, many of the ultra-new, high-tech (and *expensive*) creations of my previous analog semiconductor company were happily lapped up by, hold your breath, manufacturers of *very high-end cars.* The auto manufacturers used them knowingly to create some glittering/exciting electronics and control systems to attract buyers to their blazing new $250k MSRP convertibles. Pretty soon, the unsuspecting wannabe celebrities would drive off with the wind in their hair, without the slightest clue of all the possible things around them that were tantalizingly poised on the verge of complete failure.

2. They just haven't found out about the problem—maybe because they have a very small, underpaid, and overworked Apps/validation group. In effect they are secretly hoping the customer helps them overcome their internal human resource limitations, and helps them evaluate their creations! So if you have the desire to work for them, at least make sure you don't have a PO (purchase order) all lined up and ready to go. Take your time. Be in no big hurry to buy their product (you obviously aren't or you wouldn't be there to start with).

Question 6: *Is a particular part or component defective?* Here we are considering the possibility that the *specific* IC mounted on the board is behaving oddly for some reason. Or a *specific* component on the board. Of course we can just unsolder all the IC/components and send them en masse for vendor analysis. But that can take a very long time, and the reply is likely to be inconclusive. We need to be a little surer than that, before we go through all that trouble. So how do we identify which specific component is at fault? One of the most standard tricks in troubleshooting is to take one "good board" and one "bad board," and meticulously start swapping components from the good board to the bad one, testing after each step. We have to take extra care these days not to rip the delicate traces off, or create small inadvertent solder bridges that come back to haunt us. We should examine the board carefully after each step to ensure that it still looks good before powering up. We need to keep going doggedly in this manner till the problem goes away on the bad board. Then, the component we have just removed from the board is the culprit. That part we can test ourselves, or confidently send to the vendor for further analysis.

You can ask, does the swapping method always work? In other words, what if we go through all the components and still find that the bad board stays bad? Or the problem did go away, but the component just removed tested "good!" Puzzling! At this point we should zoom in on the PCB itself or some bench setup issue we may have overlooked. Better late than never. But once, in the process of swapping components, I believe I unknowingly cleared a very fine solder bridge, and the problem was gone. But that was deduced only from circumstantial evidence. There was no other explanation.

So is that all? Actually, there is another possibility, one that I have seen at least twice in my experience. And both are related to the magnetics.

In the first case, my colleague in Germany had observed that roughly half his AC-DC power supplies were passing the preproduction CISPR22 conducted emissions test with aplomb, whereas half were failing badly because of an inexplicable spike in the EMI spectrum. Actually that statistic itself was a vital clue, one that we all missed initially. Some other clues—all had been made in the same production batch, and the same tape-and-reels had been used throughout. None of them had any differing "histories" of any sort—none of them had, say, gone through any special testing. The engineer tried swapping some suspect components, and also the controller IC, but the problem was stubbornly stuck to the bad boards. I saw him struggle on for at least two to three weeks until he finally discovered the most unlikely cause. The EMI problem was related to the *orientation* of one of his common-mode EMI chokes. In other words, if he inserted the EMI choke with its four legs into the PCB in one way, the EMI problem showed up, but if he rotated the choke a full 180 degrees and then inserted it, the problem went away. Most EMI chokes have no silk-screened dots on them to indicate the polarity of the windings, so the chances of the production staff mounting them in one way or the other were 50:50. That is why half the boards had problems, and half didn't. We had to request the vendor of the chokes to start indicating winding polarity on future parts for us.

In the other case, the Portable Power group had just released the hysteretic switcher, the 3485. This IC ultimately sold huge quantities (via the early iPods). I happened to belong to the "evil empire"—that is, their Power Management group (though everything is relative—we thought *they* were the evil ones!). Anyway, since we all shared the same lab (and the same company in case you've forgotten), I happened to get accosted one fine day by their Apps engineer. He was very concerned why some of the boards worked just fine and some had horrible pulsing and a different switching frequency altogether. I remembered my experience in Germany immediately once he used the magic word "half." I got two hot irons, removed his inductor and flipped it around and the problem was gone. I had also noticed the feedback trace was passing just a millimeter away alongside the inductor, and on the same side. In some mysterious way, it was interacting sufficiently with the magnetic fields to throw the hysteretic controller into hysterics—but only when the orientation of the inductor was "incorrect." In this case, I recommended simply moving the feedback trace a little further away, and on to the other side of the board through the ground plane if possible. Which they did finally, and with great success. The feedback trace *always* seems to benefit from being surrounded by a sea of tranquility (i.e., the ground plane). We just have to ensure the cut in the ground plane is done judiciously, so as not to affect the natural distribution of power-related return currents (see Chapter 5).

Question 7: Is a particular part or component wrong? There are a surprising number of variations of this theme too.

1. Many years ago I remember, I had gone and bought a whole bunch of some specific CD40xx family ICs from the open marketplace for my private *garage* project (no, I didn't end up making anything close to the venerable "Mac," or even the Big Mac). Though these parts had perfectly silk-screened markings indicating a well-known Japanese brand, they were certainly not the D-type flip-flops I was expecting. I had a curve tracer built into my 20MHz low-cost Hameg scope, and it confirmed the parts were something else entirely. I don't know how that happened, but you should be aware of this possibility too. Try returning such parts to the vendor, though.

2. There was another moment of truth in the ballast project I described in Chapter 1, when we had ordered several new boards with its innovative 2N2222–2N2907 npn-pnp latch. But the protection latch on all these boards was not working at all. Design issue? We looked hard at the transistors and thought they were OK. After a couple of days we took the transistors out, finally suspecting they were faulty (maybe inadequate "hfe," etc.), and then made a discovery—we learned that so-called 2222 transistors do *not* even have the same pinouts! Each manufacturer

has its own favored pinout, even though they are all sold as equivalents, and that's how we bought them. Check that out closely, too.

3. With component sizes decreasing steadily, vendors first reached the threshold where most of us can no longer read the markings without a magnifying glass (or a Lasik procedure in my case). Thereafter, they have become so small, I just have to *presume* there are no markings at all. We just have to learn to be able to keep them apart, otherwise we will never be able to tell. But visualize an average lab bin, with small open boxes for all the components. Or even one of those small plastic shelves in a component rack, with several vertical partitions. An apparently desperate power conversion engineer grabs a few chip resistors or capacitors and accidentally/unknowingly spills/drops some of them into nearby boxes. There is now no way to tell them apart. Along comes the next engineer and picks what he or she thinks is a 10nF capacitor, but which in reality is a 0.1µF capacitor. He or she struggles with the board for weeks (even sending out data on the strange waveforms all the way from Taiwan to New Mexico), until finally suspecting a rat. The engineer then gets resigned to the arduous procedure of swapping components, and only then does he or she arrive at the doorstep of the culprit. If the engineer is very lucky, the LCR meter has not gone for calibration on that day, and he or she can finally get closure. It was the *wrong part from the right bin* all along! My preference is to keep all SMD components on their original tape-and-reels at all times, not in bins, however convenient bins may seem at first sight.

4. Inductors—be aware that many vendors put cryptic markings such as 102 or 103 on them. For capacitors, there are industry standard markings. For example, 221 is 22×10^1pf, 222 is 22×10^2pf, and so on. All are referred to the base unit, "pF." But in inductors, "102" may be 10×10^2 in *nH* or *µH*. In other words they could be a factor of 1000 apart, with the same marking. If necessary, find an LCR meter and double-check.

5. There are at least a dozen semiconductor manufacturers making the popular 384x series controllers. All of them behave slightly differently. The same applies to any other semiconductor device made by several vendors. Be cautious of so-called equivalents. So if your company's smart-alecky purchase officer has just cooked up a new deal to procure your 1N5408 diodes at half price (from some hitherto unknown manufacturer on the Mainland, for example), replace it and confirm that it is not causing the problem.

6. In one case, I remember that a standard Non-Synchronous Buck switcher IC was not working right. Everything seemed OK, the PCB, the decoupling, and so on. We tried swapping the switcher IC at first, but the problem stayed with that board. Eventually we traced it to the Schottky catch diode. We then discovered that cheap

Schottky diodes can have almost 10x the leakage current of a good Schottky. In this case, that was what was ultimately confusing the switcher IC to break into chaos every now and then. We replaced it with a quality diode and everything was fine thereafter.

Question 8: *Is there some interaction with the load?* This is a tricky one. It may involve getting to know both your prospective load and switcher IC well enough. Loads present varying profiles to the converter. Their interaction could spell trouble. A load profile is basically the *V-I* curve of the load. For example, if we have a simple resistor as the load, we know its *V-I* curve is a straight line with positive slope (compare its equation $V = IR$ with the generic equation $y = mx + b$). We can emulate this type of load by using the constant resistance (CR mode) setting on our electronic load. It is the most benign type of profile. Very few converters have any trouble starting up into a resistive load within its rated maximum. If they do, they shouldn't even be on the market!

But consider what happens if, for example, we use our DC-DC converter to power another downstream DC-DC converter. Then the downstream converter becomes our load. What is its *V-I* profile? In any switching converter, if we increase the input voltage, the input current decreases, because $V_{IN} \times I_{IN} \cong V_O \times I_O = P_{OUT}$. So this is a constant power load (down to the UVLO level). Its *V-I* profile is, geometrically speaking, a rectangular hyperbola. Alternatively expressed, if *V* increases, *I* decreases; therefore we have, in effect, a *negative* input impedance. This profile has been known to instigate severe oscillations. So at the minimum, we need to try and decouple the two DC-DC stages by placing *LC* filters between them. We can supposedly emulate this situation by using the constant power (CP mode) setting on our electronic load. But that really doesn't tell us the whole story. Because a real downstream switching converter can also send a good amount of high-frequency noise back into the upstream switching converter, causing it to delve into the exciting world of chaos. That is where the intervening *LC* filter can help. But ensure the "*L*" has very low parasitic capacitance, or it will have no blocking capability for noise frequencies.

In general, most converters are tested on the bench with the electronic load set to constant current (CC mode). True, that's not benign, nor as malignant as it gets. But the implied expectation is that converters should at least work in CC mode. They should, in particular, have no startup issues with this type of load profile. But even that may not be the end of the story! Some loads can also vary with time. For example, an incandescent bulb has a resistive profile, but its cold resistance is much lower than its hot resistance. That's why most bulbs fail towards the end of their natural lifetime just when you throw the wall switch to its ON position. And if the converter is powering a system board characterized by sudden variations in its instantaneous supply current demand, that can cause severe problems to the converter, too. The best known example of this is an AC-DC power supply inside a computer. The 12V rail goes to the hard disk, which can suddenly demand very high currents as it spins up, and then lapse back equally suddenly into a lower current mode.

These cause dynamic issues to the switching power supply, and usually the only solution to that is to have enough bulk capacitance present on the 12V output rail. Luckily, since the main feedback loop is derived from the primary 5V/3.3V rails of the power supply, there is no minimum ESR requirement for the 12V rail output capacitance, and we can freely add several electrolytic capacitors in parallel. However, modern core processors can place very fast transient load demands on the primary regulated rail, too, and for that we need a whole bunch of ceramic capacitors sitting right at the point of load. In that case we must ensure the converter is *designed* to accept ceramic loads. Otherwise it will break up into oscillations.

Remember in general, most switching converters need to be designed with no foldback of any sort. Look at the datasheet very closely for this. Otherwise they almost certainly *will* have startup problems, or recovery-from-a-fault issues, even with the CC mode setting on our electronic load. "No foldback" could mean that the overload protection present on the output rail is a simple constant current type. So, for example, a 5V/20A rail must deliver a regulated 5V until it hits 20A. Its overcurrent protection may be set at, say, 25A (to allow margin for drifts, tolerances, inaccuracies, etc.). So as soon as we hit 25A, the output voltage will start to fall, but the converter will continue to provide 25A into CC mode without any problem. Now, if we *back off* just a little, to say 24.5A, the output should immediately recover to 5V. Very few converters are that precise, however. There is some natural hysteresis involved in all current limiting circuits (and for good reason), but we are still within spec if the output comes back to 5V by the time we reduce the load to, say, 21A. But we truly have serious foldback issues if, for example, we need to reduce the load to much less than the rated maximum. We will likely see startup issues on this converter.

Another odd type of foldback is implemented in some current mode control ICs from Linear Technology. The original purpose was good—to provide good, effective current limiting under short circuits. They had realized that because of blanking time requirements, there was a minimum on-time pulse-width limitation. In other words, if there was an output short circuit, the current would hit the internal current limit of the IC and it would respond as usual, by lowering its duty cycle. But if there were a certain minimum on-time (t_{ONMIN}), corresponding to the blanking time, the controller would be unable to reduce the duty cycle beyond a certain minimum value (equal to $D_{MIN} = t_{ONMIN}/f$). In other words, the current limit is in effect not even present now! And this could cause the current to staircase above the set current limit with almost no control. In fact I was testing a similar part and found the current could go as high as 40A momentarily, for a 1.5A switcher! (I have described this current overshoot in more detail in Chapter 12). One answer to this situation is to use frequency foldback. So under fault conditions, if we lower the frequency, the minimum duty cycle becomes much smaller for a given minimum on-time. And that helps significantly in reducing the fault currents, by allowing more time for the freewheeling current to decay to zero before the next ON-pulse. But the way it was implemented in the Linear Technology chips was that the frequency of the switcher was made roughly proportional to the voltage on the feedback pin. So under a short circuit, since the voltage on the feedback pin would start collapsing, so would the frequency. It seemed simple and effective enough. But consider what happens if you are starting up naturally into the presumed maximum-rated load of the IC, and the switching frequency is too low to start with. You can then enter foldback, and you may never be able to deliver the rated maximum load of the IC at startup. But at least the relevant Linear Technology datasheets boldly carry front-page warnings that you will be able to achieve full load in CR mode, but not in CC mode.

My suggestion is to open the pdf datasheet of any prospective switcher IC and carry out a text search (Ctrl + F) for the world foldback. If you find it, question the vendor about its full impact before you select the part for your application. Foldback is, in general, a good idea in terms of protecting the converter under abnormal conditions, but it should be used very judiciously so as not to impact *normal* behavior. For example, the Simple Switcher family has a hidden second-level current limit protection at which frequency foldback (or skipped pulses) occurs. But that trip level can only be encountered under very severe conditions—namely, a sudden overload with a completely incorrectly sized inductor that hard-saturates in the process. At other times it is not encountered and doesn't therefore interfere. It is considered transparent to all but the *most novice* engineers. And that is what I consider the right type of foldback.

An exception is the foldback behavior discovered in the third-gen 267x family whenever the *duty cycle exceeds 50%*. That has really nothing to do with the protection of the IC, though it can be successfully argued that this belatedly discovered "feature" does eventually help in that respect—by almost turning off the IC altogether (yes, that would work!). Read the following clarification apparently issued by them on their public discussion forum. Keep a few bags of salt readily available.

> *March 1, 2006:* The condition described is the result of what amounts to a foldback current limit design that's intended to prevent damage to either the regulator or the load under unusual fault conditions. Anyone familiar with foldback current limit will realize that there are always conditions that can be realized that force the foldback to get "stuck" in a stable, low output voltage operating mode. The solution in general is to reduce the load until the output is allowed to recover. The datasheet clearly advises the user what to look for and how to deal with potential problems that may arise from this. Any implication that the information is deliberately obscured is clearly misleading. . . .
>
> National Semiconductor, 2006, Discussion Forum response,
> http://wwwd.national.com/national/PowerMB.nsf/

Page 12 is clearly "non-obscure" from now on. Also, what this doesn't explain is why the protection activates only above 50% duty cycle, and why all previous and subsequent switchers from the same company (and all others) did not and do not have this type of intrusive foldback. They also all have a second-level foldback current limit anyway, so why does this particular device family need *two* foldback circuits? It also does not tell you that it is not about just "reducing the load and allowing the output to recover" but that this IC fails to even *start up* at *half* the rated current, if you have innocently set, say, 12V output from an 18V input. But surprisingly, if you have set 12V input and 5V output, for some mysterious reason the company now thinks you don't need any further foldback protection, and hey presto, your startup is perfect. I couldn't put up an engineering rejoinder on their discussion forum, because by then they had thoughtfully deleted my login privileges! In other words, they decided I was the problem (with their chip) all along. And that I finally needed to remain equally obscure from now on.

Question 9: *Is there some interaction with nearby circuitry?* Yes, you could be picking up fields from nearby circuits, but that shouldn't affect a typical switcher, simply because it produces enough noise and fields of its own. However, it is a good idea to do the reverse-peel here. If I find the converter is on a larger system board, I immediately and carefully first cut off all the traces leading from its output and divert them to my predictable electronic load. I also cut the input traces and divert them to my bench power supply. If the problem is gone, it is an interaction problem.

One of the most obvious mistakes customers make is to try and parallel several DC-DC converters off the same input. They make the situation worse by allocating one full layer of their board to the supply Vcc. I always like to see a nice ground plane, but in such cases I would consider creating *two* big ground islands, one under each converter, and then connect them together at a single point so as to avoid interactions. But the worst thing you can do is to have the two converters share a complete ground plane *and a full input supply plane* too. Basically, the two converters are no longer independent because there are no intervening trace impedances between them. Take a look at the upper section of Figure 8-2 and tell me why converter A won't draw its input current from the input capacitor supposedly assigned to converter B. The schematic is, incidentally, again lying, though in the opposite sense now. It may be making you think each converter is separate from the other, whereas in reality they are not. They *will* therefore interact, and it is impossible to predict how this will affect their performance. My preferred layout is to create long thin traces going to their respective Vcc (i.e., V_{IN}) pins as shown in the lower section of Figure 8-2. That way the converters do not interact much, though a more formal solution is to insert small LC input filters. We have to be very careful, however, of not introducing any significant "*L*" to the input side of any DC-DC converter, because this affects the ability of its input voltage source to refresh its decoupling capacitors quickly enough, and so the wobble on its input pin can increase sufficiently and trigger oscillations or chaotic behavior on its own. Also, even little beads on the input of a Buck or a Buck-Boost have been known to generate nasty inductive spikes of their own, which can kill the IC eventually.

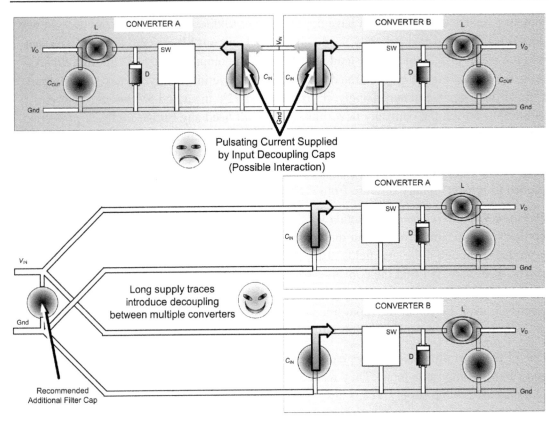

Figure 8-2 How to Run Multiple Converters from the Same Supply

Question 10: *Am I trying to achieve something that is really possible?* The number of young engineers who try to parallel converters for higher power and then fall flat on their faces is legendary. Well, you can't just take two switchers and tie their outputs together for higher power. The problem is that they have very high gain error amplifier stages and their reference voltages are not exactly the same. For example, if one of them has a reference voltage of 2.5V and the other is at 2.51V, and suppose they have perfect 10kΩ resistors on their respective voltage dividers, we expect 5.00V output on one converter and 5.02V on the other. If you think they will settle down nicely at either an average of 5.01V, you are wrong. Because if that were so, one converter's feedback pin would still be 5mV below its reference level, whereas the other would be 5mV above its reference. And that will cause one converter to go to almost max duty cycle in an effort to bring the output voltage up, whereas the other will go to near-minimum duty cycle to bring it down. But for sure, there is no synchronized teamwork in the works here! They will end up fighting with each other and if you are lucky, the output will stabilize at some intermediate/average level. But if you measure the *currents* through each of the two inductors, you will find the net load current is far from being shared equally. So two 2A switchers won't give you anything close

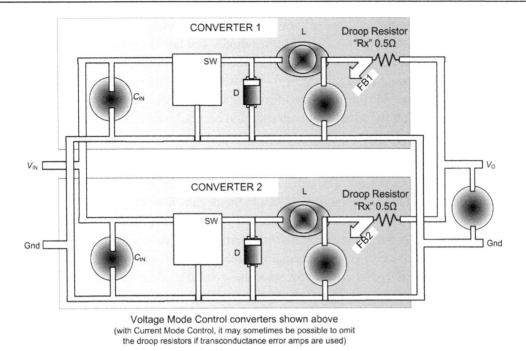

Voltage Mode Control converters shown above
(with Current Mode Control, it may sometimes be possible to omit
the droop resistors if transconductance error amps are used)

Figure 8-3 Paralleling Converters for Higher Power

to 4A, because the current limit of one or the other will activate (and then they will start motorboating). You may be able to get say 3A, but it could end up being distributed as 2A and 1A, certainly not 1.5A each. It is often also said rather blithely that "current-mode switcher ICs can be easily paralleled." But try it and you will discover that too is not possible, at least in such a simple manner. Don't ever attempt the impossible. Ignore the marketers. To succeed here, you will either need fairly large ballasting resistors on each of the outputs or a dedicated load-share IC (as from Unitrode/TI). If you try the ballasting technique shown in Figure 8-3, you need to carefully calculate what value of resistance you need. Note that the feedback to each converter needs to be taken from the *left* side of these resistors. Also, if you use too large a resistance, the current sharing will tend to get better and better, but obviously the output will droop dramatically. At best you can search for a good compromise.

Another milder example of this is a standard Non-Synchronous Buck switcher IC. Every single "typical applications" diagram on the datasheet shows a Schottky diode, without perhaps explicitly stating as much. This is an example of an "implied expectation" on the part of the *vendor*—that *you*, the customer, won't miss the *truly obvious*. Yet there are many who think they have achieved some slender advantage in substituting an ultra-fast diode in its place. First read the Abs Max section of the EC tables carefully. Most vendors specify that the SW node should never be taken more than 0.4V below IC ground. That is because they expect substrate currents flowing back into the chip, affecting its performance and possibly damaging it. So if you use an ultra-fast diode, you are almost certainly forcing the SW node roughly 1 to 1.5V below ground when the switch turns OFF (equal to the forward voltage across the diode). You are on your own now. Incidentally, while doing a survey on this topic, I learned that Maxim Integrated Products typically specifies a maximum of only 0.3V below ground (hardly achievable with even the best Schottky diode in the universe), whereas Linear Technology doesn't seem to even want to specify this Abs Max parameter in most of their datasheets.

Incidentally, there are customers who come and ask, "I know you have stated in your Abs Max table that I shouldn't apply more than 24V to the device. But what if I apply 28V for just 1ms?" The *principled* answer to that is, you can't apply even 24.01V, for even 10^{-12} seconds! The company officially doesn't stand by it. Yes, internally they do test at higher stress levels than published, and have also got various guard-bands present (for their protection and reputation). But remember you don't know what these are. Also, keep in mind that voltage overstress leads to almost instantaneous death, whereas current ratings are related more to internal heat buildup, so you can always exceed them somewhat for a short time.

Question 11: *Am I trying to achieve something that is generally known to be risky?* Yes, if you are trying to use an SCR crowbar on the output for overvoltage protection, for example. There is enough industry experience by now that these can trigger spuriously and should be avoided. Rather than troubleshoot this, *replace it* quickly. You may get a few prototypes working satisfactorily on the bench, but do a mass production on this, and your Boss will certainly overvoltage and lock you out.

Similarly, if you are trying to use current-mode control for your *Half-Bridge*, you should know that that control method is well-suited for a Push-Pull topology, for example, but it actually aggravates the chances of flux staircasing and core saturation in a Half-Bridge. Oh yes, you should also know that voltage mode control will *not* protect the *Push-Pull*. There you need current mode control. By the way, who makes a Push-Pull with voltage mode control nowadays? Try the 5033!

I also remember years ago, my colleague was struggling with the Push-Pull topology for his high-power inverter project (yes we had three engineers working simultaneously on inverters at that time, and we all learned *what not to do*!). We were all using the popular voltage-mode 3524 controller IC at that time. My colleague was achieving *great success* (everything is relative). He could actually run it at 500W for about 10 minutes, and then it would explode with a huge bang, opening up all the high-current circuit breakers he had thoughtfully put in series with it. One evening he was getting extremely puzzled and called me to show me something. He had just noticed that in the few minutes preceding the blow up, the waveform of the Push-Pull would develop a mysterious edge as shown in Figure 8-4. But only for one transistor! I never fully figured this out for years. Now I realize

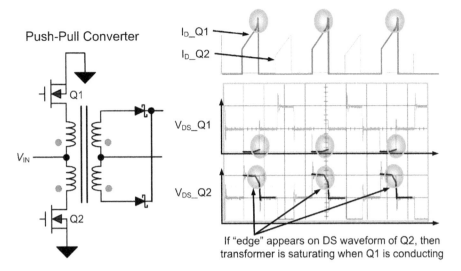

Figure 8-4 Signs of Core Saturation in the Push-Pull, with Voltage-Mode Control

that the core had staircased to one side of the BH curve (core imbalance), and was saturating at the point where the edge was. So, essentially, the core had lost its ability to hold any voltage across it at that moment. We didn't have a current probe those days and probably hadn't put in a sense resistor either. If we had *monitored the currents*, we would have seen the cause. We had tried larger cores too, but I am now convinced that almost no core can ultimately prevent this slow creeping death (staircasing) in a Push-Pull with voltage-mode control. You could depend on good current limiting to save the switches (but not necessarily to save the performance of the converter), because the truth is the core is running completely imbalanced, and so are the two "halves" of the Push-Pull converter (the two switches, the two winding halves, the output diodes, etc.). The only reasonable way out is to move to peak current mode control for this topology. A contributory factor to our early disasters perhaps was the fact that current limiting on these early devices was not really effective. Today it is common to design any IC such that if the current limit is ever reached, the switch is turned OFF firmly for the *entire duration* of that switching cycle (latched). But these ICs would turn the switch OFF when the current limit was reached, but as soon as the current dipped below the current limit threshold, the switch would turn ON again. So it would sort off buzz away around the current limit region, eventually causing enough noise to break through completely and damage the switch. I believe they fixed it later. I also checked that, as of today, Texas Instruments still sells the 3524 (accompanied by its vintage 1977 datasheet, last revised in 2003), and also includes a typical schematic for Push-Pull applications. Think about it—Unitrode (now part of TI) were the original pioneers of current mode control and heavily publicized all the above-mentioned weaknesses of 3524-based, TL494-based (voltage-mode control) Push-Pull topologies.

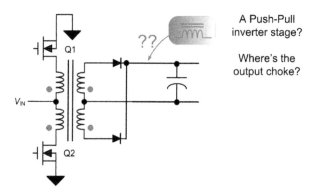

**Figure 8-5 Low-frequency Inverter Designs Do
Not Beget High-frequency Switchers**

As mentioned briefly, a recent contender to the hall of fame is the 5033, officially labeled a "100V Push-Pull Voltage-Mode PWM controller," and released in 2003. Luckily, this analog vendor's datasheet only shows a Half-Bridge at work, and that we are aware is a good match for voltage-mode control. So I personally tend to think that marketing (alone) was responsible for this misleading push (or pull).

Another inverter my colleague was making years ago looked a lot like Figure 8-5. He had been having some success, and was feeling optimistic, until I asked him *where the output choke was*! You don't make a Forward converter without an output choke! He had apparently been lured astray by similar looking schematics of traditional AC inverters made from iron laminations. But this was a high-frequency switcher, man!

Any Buck-derived topology (e.g., the Forward converter, the Half-Bridge, the Push-Pull, the Full-Bridge, etc.) needs an output choke. Otherwise it is akin to running a Buck *without its inductor*—you can thereby create a dead short cross the input supply rails.

Another common mistake we used to make in those days, and one that a very large number of engineers still make, is that if we ever thought the transformer might be getting too close to saturation, we would quickly wind another bobbin with *additional* turns on it. We thought we should increase the inductance and thereby reduce the peak currents, and that would help. Actually, this intuition is probably again a leftover of the days of winding big AC line transformers with CRGO laminations. Those were almost impossible to saturate, which is why in their design manuals, vendors would often give you an equation to calculate N, which was the *minimum* number of turns of the Primary. In switchers, the picture changes entirely, because if your transformer (or inductor) is saturating, you actually need to *reduce* the number of turns (and *reduce* the inductance). You are puzzled, because you are mentally thinking that the peak currents would then be even higher, and so the chances of saturating your transformer would be greater. Wrong! The reason for saturation is not I_{PK} alone, but $1/2 \times L \times I_{PK}^2$. That determines the energy-handling capability of any core. So suppose your inductor is designed for a 1A with ±20% current ripple, and you double the number of turns. Yes, your peak current will decrease. By how much? Remember that inductance is inversely proportional to the ΔI, which in this case was 0.4A to start with. So by doubling the number of turns, you have increased the inductance four times (L being proportional to N^2), and the ΔI therefore goes from 0.4A to 0.1A. So now your peak current is $1A + (0.1/2)$ = 1.05A. Which is about 1A. But now calculate the product $1/2 \times L \times I_{PK}^2$. This has gone up almost four times because of the increase in inductance (with very little corresponding reduction in the peak current). That simply means you need a transformer/inductor about four times bigger now. So how do you ever expect to solve the problem of core saturation by *increasing* the number of turns? You must always keep in mind that in switchers, *smaller inductance leads to smaller inductors.* Converters that use DCM or BCM (boundary conduction mode) will always feature much smaller magnetic components than those operated in CCM. Their inductance is much smaller! The problem with them is that *other* components may need to be unnecessarily larger, such as the switch and input/output capacitors. See Chapter 12, too!

Another IC designer's dream that can sour quickly is a Flyback with 100% duty cycle. We all know that a Flyback delivers power to the output only when the switch turns OFF. But if you have 100% duty cycle to start with, there will be no energy going to the output, so the feedback pin would remain at zero, and the controller would never know it now needs to start pulling back on the duty cycle. The switch could stay on forever in an effort to get the output to rise. You can easily get into a self-destructive Catch-22 situation here, if the current limit and/or soft-start do not step in quickly enough to save the show. Maybe that is why I could not find a single Flyback IC out there with 100% maximum duty cycle. You will find plenty of Buck ICs with 100% max duty cycle, but not Buck-Boost (i.e., Boost) ICs, though an exception that proves the rule is the 3478/3488. Judging by the datasheets' front pages, these devices are *somehow* intended to work beautifully for all Flyback, Boost, and Sepic applications. I doubt that. But they do have "proof" in the form of an online seminar called "Designing DC-DC Power Supplies using High Performance Switching Controllers." In that 2001 product release collateral, we see a young, motivated engineer enthusiastically delivering a message of excellence vis-à-vis these specific products, with the legend of Analog (the self-proclaimed "Czar of the Bandgap") sitting right beside him in full regalia. I would really like to personally believe that the king *wasn't* there vouching for these products. Because he does end up giving an aura of credence to these ICs, very undeserved in my opinion.

I would recommend you try nothing overtly risky in power, especially not by ignoring well-known industry experiences. There could be a high price to pay, and troubleshooting the boards could be the very least of your burgeoning problems. *"Power" hates to be taken for granted*, as we all discover sooner or later. Also carefully go through discussion forums to see what problems others may be facing with the proposed part. Don't fall for the possibly glib/evasive company responses, though. Just *count* the queries and that should tell you. Also don't forget to read Chapter 12.

Question 12: *Am I the one somehow managing to inflict damage on the IC?* If you ever suspect the IC is damaged, just pack it in an ESD bag and send it off for failure analysis. But then suppose the next part looks good to start with and eventually develops similar symptoms. You should then realize you are somehow managing to damage the part, without realizing it. Here are some interesting (and common) examples of this.

    a) You have a regular Synchronous low-voltage Buck switcher working on your bench. You unplug the load, power down, then decide to immediately power up again for some reason. The IC gets destroyed almost immediately. You send it for failure analysis and a few days later they tell you the part was damaged by a high voltage on the input pin. Why?

        When you decreased the load to zero, the IC probably entered energy saving mode (PFM), in which the lower transistor turns OFF permanently. So the output capacitor stayed almost fully charged up when you powered down and powered up again. But when your converter tried to start up again, it did so with its usual "soft-start." Therefore the duty cycle was very low to start with, and the lower Fet stayed ON for most of the time each cycle (normal Synchronous/complementary drive). But because the output capacitor was almost fully charged, it drove a huge current in reverse direction through the inductor (see path 1 in Figure 8-6). At some point the lower Fet turned OFF—and all this reverse inductor current cycled into the input capacitor (see path 2 in Figure 8-6). If the high-side Fet were ON, the current went through its channel, but if the Fet were OFF, the current went through its body diode. Either way, all the output energy starts getting dumped into the input capacitor, raising the voltage on the input pin. Basically, what has happened is that the Buck switcher has momentarily become a Boost switcher in the opposite direction! To avoid this situation, you may need to pick an IC that is designed specifically to handle such pre-biased load conditions and/or to increase the input bulk capacitor significantly. One of the ways to do this is shown in the lowermost part of Figure 8-6—basically, we need to implement complementary soft-start for the lower Fet, too.

    b) You have an older generation part with an external voltage divider. Since you want to use it with a ceramic capacitor on the output, you have thoughtfully put in a feedforward capacitor across the upper resistor of the divider. But then you short the output a few times and the part gets damaged. Why?

        The feedforward capacitor C_{FF} shown in Figure 8-7 has a voltage across it in steady state. When you short the output, the feedforward capacitor cannot

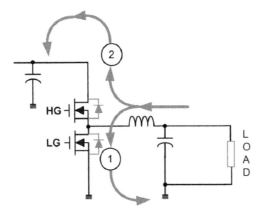

Start-up Waveforms

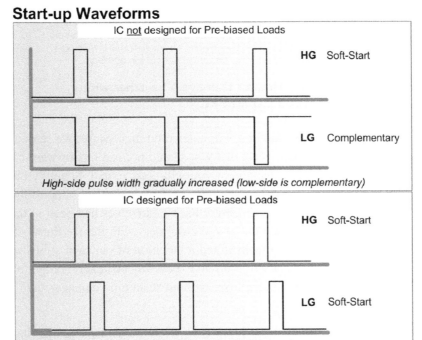

Figure 8-6 How a Buck Turns into a Boost in Pre-biased Load Conditions

discharge immediately, so its lower end gets pushed below ground (the capacitor holds the voltage across it for some time). This eventually causes an unexpected current passing through the ESD diode present at the feedback pin of the IC, in effect a sneak discharge path for the feedforward capacitor. Feedback pins are almost invariably not allowed to go more than 0.3 or 0.4V below IC ground to prevent such damage. Therefore, a few years ago, as soon

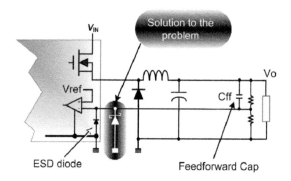

Cff develops a voltage equal to Vo-Vref in steady state. When the output is shorted, its upper end is dragged to zero. Since it cannot discharge immediately, its lower end tries to go to -(Vo-Vref). But the ESD diode then conducts causing damage to the IC. The energy in Cff largely gets dissipated in this ESD diode. If Cff is large, an alternative path must be provided - a small-signal, low-voltage external Schottky between Feeback pin and Gnd.

Figure 8-7 How to Damage a Switcher with a Feedforward Capacitor C_{FF}

as I discovered these failures and understood their cause, we started specifically mentioning in the datasheets that you should not use feedforward capacitors larger than a certain value and/or you will need a small Schottky diode from feedback pin to ground. My battle-honed Boss, who had hitherto thought he had seen it all, was quite surprised that we ourselves had been in the position of recommending typical circuits to customers with the feedforward capacitor present, little realizing it constituted a violation of our own published Abs Max ratings on the Feedback Pin (you do expect any switcher's output to be shorted and released in its normal course, without sustaining damage—an implied expectation though).

c) Your bench power supply is set fairly close to the maximum input voltage rating of your IC. You have just changed the input ceramic capacitor of the converter from 22μF to 10μF, and the supply line still looks very clean (under steady conditions). But the part gets damaged almost every time you connect the red lead coming from your bench power supply directly to your board. Why?

The long inductances of the leads, combined with the low-ESR input capacitor and the negative input impedance of a switching converter, can produce a lethal undamped oscillatory circuit that can produce huge input swings, often exceeding the ratings of the IC. There are two ways out—either try to use only high-ESR capacitors at the input (or at least parallel a high-ESR electrolytic with the ceramic input capacitor), or increase the amount of bulk capacitance.

Therefore, a 22μF input ceramic will give a smaller input overshoot than a 10μF input ceramic. A 47μF will give even less, and so on.

Note you will not see this failure mode if you first plug in your converter to the bench power supply and then turn on the supply. Because, in that case, the output of the supply comes up very benignly as it first charges up the hefty bulk capacitors sitting inside it across its output terminals. The only way to instigate this wild input overshoot is to jam the banana plug into an *already powered-up* bench power supply. This produces the highest d*V*/d*t* possible at the inputs of the converter. Further, this "hard d*V*/d*t* test" is not only a tool to see the input overshoot, but it is a very good diagnostic tool in general for exposing any latent weakness in the IC. I do this almost invariably during testing. There are always surprises in store! Often, this alone can call for a significant increase in the input bulk capacitor.

d) You have a Non-Synchronous Buck switcher IC powering a load. You reduce the load to zero and then attempt to discharge the input capacitors of the converter. The IC gets damaged. Why?

    If the switch is a Fet, a momentary surge current will flow from the output capacitors through its body diode, discharging the output capacitor. The device is not usually tested by semiconductor manufacturers in this mode, but neither has there been much evidence of reported field failures in this manner, unless of course the output bulk capacitance is very large and/or it is charged to a high voltage (energy in a capacitor is $1/2 \times C \times V^2$).

    If the switch is a BJT, this is a clear no-no because a bipolar attempts to block reverse voltage, but is really not designed to operate with any reverse collector-emitter voltage.

e) You have set up a Buck switcher IC with a BJT switch to deliver constant current. You intend to use it to charge a battery connected directly cross the output terminals of the converter. But you end up constantly destroying your switcher IC. Why?

    For the same reasons above. Think of a battery as an infinitely large capacitor. The only way to handle battery charging with a BJT switch is to put in a blocking diode in series with the battery.

f) You have a Boost IC set to deliver 12V @ 1A from 5V. You run a spreadsheet, which suggests you use a 4.7μH inductor. So you pick a 4.7μH/1.5A inductor from the bin. But the IC fails. Why?

    The average inductor current in a Buck delivering a load current of I_O is I_O. But in a Boost or Buck-Boost, the average inductor current is equal to $I_O/(1 - D)$. Further, the peak current in all cases is typically 20 to 30% higher than the average inductor current (by the normal selection criterion for inductance). We have to calculate the worst-case peak value and use it as the minimum rating of the inductor.

g) In an effort to improve the efficiency of your Buck design, you have picked a Synchronous Buck controller IC simply because it has very high-current drivers. But both the Fets blow up every now and then. Why?

Be very careful of overly *aggressive* drivers. Such ICs can damage themselves in several ways. In general the fast transitions can induce spikes all over the board, causing weird problems everywhere, including general controller malfunction. But in modern Synchronous Buck converters, one of the strong reasons for *slowing* down the Fets and picking Fets more carefully is the phenomenon of "CdV/dt turn-on." If you look closely at the gate of the lower Fet (when using a controller IC), you will see a small blip on it the moment the high-side Fet turns ON. In effect both high-side and low-side Fets are briefly on *simultaneously*. What is happening here is that at the moment the high-side Fet turns ON, it pulls up the SW node very dramatically. This changing voltage induces a small current to flow through the Drain-to-Gate capacitance of the Fet (as per $I = CdV/dt$), and this can turn the lower Fet ON. Eventually, this can provoke cross-conduction, which will either be totally destructive or, at the bare minimum, will lead to a substantial loss in efficiency. That efficiency hit becomes especially noticeable when the converter is in normal Synchronous mode (forced PWM mode, *not* cycle skipping mode) at very light loads. My usual test is to benchmark the zero-load supply current for a good board, and then I can easily detect excessive cross-conduction if I see more than a few mA in excess of that level. If this is a controller IC, I also like to compare prospective low-side Fets during the initial selection process, in terms of the ratio C_{GD}/C_{GS} (equivalently C_{RSS}/C_{ISS}). A lower ratio makes the Fet less susceptible, and similarly, a slightly higher threshold voltage V_T improves the Fet's immunity against this spurious turn-on effect. In one IC design situation a few years ago, we actually ended up "rev'ing the silicon" one last time just to make the drivers far "less aggressive." The pull-up was reduced by at least half, to slow down the turn-on. And that also saved significant silicon area and led to a better product in general.

So, if you can access the gates of the Fets, try putting in small resistors in series with them. If it is a switcher IC (with no access to the gates), try inserting a small resistor in series with the decoupling capacitor of the driver supply (usually a 0.1μF capacitor attached to the Vdd pin and/or the bootstrap pin). Better still, pick an IC with less aggressive drive to start with. Because otherwise it *will* commit suicide sooner or later.

Effective Bench Work

Introduction

As I said, I first need to emphasize underlying principles rather strongly. Now, with that behind us, we are in a much better position to understand this particular chapter.

Basic Equipment

This book is not a substitute for an equipment manual! You must familiarize yourself well with the details of your specific instruments.

The most important piece of equipment on your bench is the oscilloscope. In Chapter 7, I mentioned an incident plucked straight from the annals of that timeless battle between analog scopes and digital scopes. It should help us all realize that though analog scopes may not be able to see something that is non-repetitive or too brief, digitals can very easily miss something that *does* exist. Yes, in digital scopes you can also end up seeing something that *doesn't exist*—they call it *aliasing*. I just say, "*check you are really on μs/div, not ms/div!*" For this reason I tend to agree when they say, "analogs don't lie." Yes, they may not tell you *everything* on occasion, but at least you usually know beforehand what that missing information is likely to be. On the other hand, digitals have been caught virtually lying through their teeth on occasion, besides simply not telling you the whole story. And worse, with digitals you may never know what you have missed, until of course those old-fashioned guys sitting around their analog scopes on Ubi Avenue (in Singapore), *tell* you *what* to start looking for. However, to be fair to the digital era, they do say nowadays digitals have become so good, they *almost* do everything an analog could do, besides a lot more of course (based on their unique storage capabilities, naturally). But to exploit that capability, unlike an analog scope, you often have to be *already* looking for something specific in mind, almost know *beforehand* what it is, and then *set your digital scope carefully* to capture that event. Its controls are not easy, and setting them *appropriately* can get very tricky at times and calls for a lot of bench experience.

Lab Essentials

Here are some things to remember:

a) Make sure your probes are well compensated. You might see an overshoot that *doesn't* exist, if you don't! Or a soft-start that the designer isn't even aware of.

b) Make sure you have, for example, Channel 4 set on 50Ω input impedance and a vertical scale of 10mV/div, and then connect it to the current probe amplifier (e.g., from Tektronix). Or declare that the current limit has somehow almost doubled.

c) Degauss your current probe. Note that if it doesn't detect the right impedance at your scope end (i.e., 50Ω), it usually won't complete this task. And you may or may not notice that its digits are flashing very differently just to alert you.

d) Make sure you set all the remaining three channels to the 1MΩ setting, and your probe tips on the 10:1 mode. Note that for a noise reading, the 1:1 mode is usually considered more advisable. The reason for that is when you set your probe to 10:1, you are basically using an internal divider in the probe tip to reduce the picked up signal by a factor of 10. The prime advantage of doing this is that the input impedance of the probe tip also falls very low. In fact, it goes from about 1MΩ and 30pF in parallel in 1:1 mode, to about 10MΩ with 3pF in parallel in 10:1 mode. So the measurement becomes less invasive. But with this setting, you also end up worsening the signal-to-noise ratio. In particular, when the scope increases its gain automatically (to compensate for the 10:1 mode), it ends up bringing up its own noise floor too. So the 1:1 mode is less noisy, inherently so. Unfortunately the probe tip capacitance of 30pF can create its own problems and can often quell the very noise you are trying to measure.

e) Keep in mind, however, that this 3pF or 30pF of tip capacitance may be a useful *diagnostic tool* on occasion. I often go to the component bin and pick out a 22pF capacitor, just to place between two points to see if a certain suspected noise pickup goes away. Sometimes, if one of those two points is the ground, the scope probe tip can suffice! For example, if you are trying to trace the route of some stubborn noise into the IC, you will suddenly realize which pin may be the culprit, simply by putting a probe tip on it. The capacitance will usually kill the noise (not the converter), and everything will work nicely thereafter (so long as the tip is kept pressed on!). This "probe-touch" technique can help identify noise-sensitive nodes in general, such as current sense pins, and so on, though I have usually had bad luck trying to touch a probe tip to the feedback pin.

f) Remove all bandwidth limiting on the scopes to start with. Very rarely do you want to miss out any vital information about your power supply by limiting the scope bandwidth. The only time you may want to do that is you are sure there is too much *extraneous* noise. An example of this would be an output *ripple* measurement, where you would naturally want to suppress the noise component. Of course, never try to look at the *noise* component using bandwidth limiting! Ways to do a proper Noise and Ripple measurement have been discussed extensively in Chapter 3. Also, don't forget the grounded probe test technique mentioned there.

g) Cut out a suitable place on the PCB to insert a DC current probe. In DC-DC converters, this should be in series with the inductor, *nowhere else*. In AC-DC converters, you may want to slip the probe in series with the Drain of the Fet (*not the Source!*). But in AC-DC converters, you can also monitor the switch current just by connecting a (voltage) probe across the sense resistor (between Source and Primary Ground).

h) Note that in AC-DC converters, you might like to *not* connect the AC until the rest of the circuit looks OK. So you may need to connect a separate DC supply rail to power up the control IC first. The problem with current-mode control ICs such as the 3842 is that if you don't connect AC power, you don't have any current sense signal either, so there is no ramp available at the input of the PWM comparator inside the IC. Your output pin will thus stay either high or low permanently, with no switching to confirm all is fine. So, to derive more meaningful results at this preliminary stage, you can try injecting some (more) of the clock ramp onto the current sense pin to "fool" the IC into thinking there is some switch current flowing. This you can do by simply increasing the slope compensation capacitor that I spoke about in Chapter 1 (Figure 1-4). This is strictly a temporary measure, of course. Also remember, the error amplifier output is available on Pin 1 of this IC (labeled COMP), so by changing the voltage on that, you can get some realistic switching and duty cycle variation out of the IC, even before you connect AC power. Note that you should *not* use two error amplifiers. So if you have a TL431 on the Secondary side, for example, you should deactivate the error amplifier of the 3842 completely, by pulling Pin 2 to ground. The opto-coupler should then be directly connected between COMP and ground. I usually prefer to connect a very low AC or DC voltage to the input of my supply, and put in a fairly large current sense resistor initially, so I could get a strong sense signal from it. Just to be extra safe, the connection to the low-voltage input AC (or DC) should be current limited in some way. I would use either a typical DC lab power supply at the input of the power supply, or set up two cascaded variacs for providing a very low-voltage AC (at the same point). Though in the latter case, I would often also insert one or two standard 60W to 100W incandescent household bulbs (lamps) in series, between the output of the variac and the input of my AC-DC supply. That is a big help especially when your switch fails—as you will see (and brightly so)!

i) Set up a good electronic load *in CC mode*. Remember Question 8 in Chapter 8, "Is there some strange interaction with the load?" Remember that some of these can display unexpected glitches sometimes. So just be watchful, not suspicious. My favorite loads are ones from HP. In Singapore we used to use "Prodigit" loads (we thought they were fairly good and quite cheap at that time). In Germany we bought a whole bunch of Prodigits (on my recommendation), but before that they were using mainly Chroma or Kikusui loads. Of course these are just suggestions, I am sure there are many loads out there that may be a better choice for a certain application. For the 400V rail of a PFC stage, I have used incandescent lamps as load.

Clock Instability and Jitter

Check the *clock*. In DC-DC converters you can simply look carefully at the SW node. It should be stable; otherwise please read up some more on PCB routing and input decoupling in previous chapters, and then return to this point. When designing AC-DC converters, I used to look very hard at the signal coming out of the IC meant for driving the Gate of the switch (Pin called OUT on the 3842).

So, the first goal is to ensure that you have a *stable* clock, running at the *correct frequency*. You should also check the clock *and* the actual switching (or driver output) waveform once you have powered up completely. When you finally do have a *regulated* output, *one of the edges* of your switching waveform will have a small amount of jitter—because that's exactly *how it is regulating*. A certain amount of jitter is acceptable and natural. In Figure 9-1 this is the blurred *falling* edge of the waveform marked OUTPUT. In a 3842, the OUTPUT pin is the drive to the Gate of the N-Fet. *If you have a certain amount of noise or instability, the jitter will be too high. If you have severe noise, your clock will be unstable. You need to fix your clock before you take up the jitter.* Take a look at the lower half of Figure 9-1 and see how the final "Output" waveform becomes unpredictable because of the noise. What is happening here is that as the Output waveform goes low (i.e., Switch turning OFF), the noise associated with that transition edge gets fed through into the clock circuitry and manages to terminate the Clock, too (it too goes low *prematurely*, and then starts afresh with the next cycle, *but ahead of schedule!*). The *frequency* of the switcher will be seen to be varying, and not with any regularity, either. Remember that under normal operation, the clock is supposed to "time out" a little *after* the Output goes low. In other words, *the clock is supposed to determine the output, not the other way around.*

What are the reasons for clock instability? High-frequency noise is always generated at turn-on and turnoff in any switcher. This noise can infiltrate into the IC via various pins. It can be very hard to filter out and control. You may need to ultimately simply avoid turning the Fet OFF too dramatically. In most switchers, the turn-on transition is traditionally delayed (or slowed) just a little, so as to allow the output/catch diodes to recover

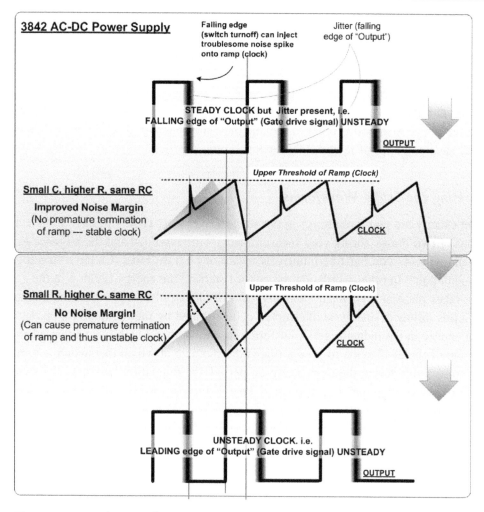

Figure 9-1 Understanding the Difference Between Jitter and Clock Instability

sufficiently before the Fet turns ON. Otherwise a huge (reverse recovery) current spike can flow through the diode (and Fet) during the turn-on transition, seriously undermining the efficiency. But many engineers tend to instinctively state that "turn-on should be slow, turnoff should be as fast as possible." Not really so. As you can see, that can sometimes throw the clock completely off-balance, possibly leading to immediate switch failure. That is the reason why, when you start troubleshooting, *start with fairly large Gate turn-on and turnoff resistors*, before you try to optimize the entire performance of the converter. It is important to *first* get the supply to switch *reliably*, then ensure that all the current limits, duty cycle clamps, line feedforward sections, and so on, are working nicely. Only *then* should you make that last burst for efficiency. In other words, *learn to prioritize tasks when debugging*. Realize solemnly that causing switch failure is not going to help you check out anything other than the patience of your prototype production staff. One thing *you* can do is to *improve the noise margins* of your circuit/control IC. One stratagem for that is indicated in Figure 9-1 itself. You can see how different RC combinations can affect the noise headroom, that is, the voltage between the tip of the noise spike and the upper threshold level of the ramp. Also don't forget to include the Gate zener I talked about in Chapter 1. That will save you a lot of rework (and unwelcoming scowls).

Interpreting the Scope Waveforms

So, what exactly are you looking for in the scope waveform? Here you have to be actually conscious of two things during your measurements—the *topology* and the *triggering*. Let me explain why. As you must be intuitively aware of by now, most ICs have "trailing edge modulation." In other words, the process of starting the energy buildup in the inductor takes place at *predefined moments* (start of clock pulse). However, the moment at which this inflow of energy is interrupted is determined by the regulation loop, and then the energy in the inductor gets transferred to the output. Which means that when the clock ramp of the 3842 starts to go up, that is the moment at which the switch is turned ON. At some point during the clock ramp-up period, the regulation loop asks the switch output to basically "cop out" because "hey! That is enough for now, the output is up there already." So the switch is turned OFF, *but the clock continues its ramp up till it finishes the cycle*. After reaching the upper threshold, the clock ramps down very quickly to the lower threshold (the ramp down period being the minimum off-time in the 3842). Thereafter the whole sequence starts again. However, you have to be very careful which waveform and which edge of it should be used to check the jitter and the clock, *and that depends on the topology*. For example, if you are looking at the OUTPUT pin of your 3842, you must set your scope to trigger on the rising edge. If you are looking at the Drain waveform, set it to trigger on the falling edge. If it is a Buck converter, trigger on the rising edge of the SW node. If it is a Boost, trigger on the falling edge of the SW node. *Basically, for any topology, always first ask yourself, "which edge corresponds to energy starting to get delivered into the inductor?" And that's always the edge you want to trigger your scope on*. In general, you will then see something similar to the square waveforms of Figure 9-1 (or an inverted version of that). Specifically, look at the examples in Figure 9-2 and Figure 9-3, and see how the triggering of your scope needs to be set depending on the topology, and how to interpret the waveforms as being normal or abnormal. Note that the fuzz on the edge used to be very easy to see on an analog scope. On a digital scope such as the TDS420/460 series, you may want to set it to display the waveform on "average" setting (maybe 10–20 waveforms superimposed), instead of the "high-res" or "sample" capture modes. It will create a fuzz similar to an analog scope.

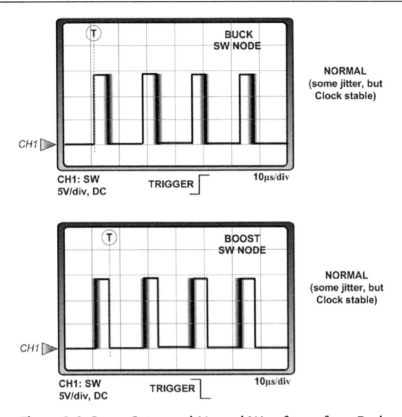

Figure 9-2 Scope Setup and Normal Waveforms for a Buck and a Boost

Now you can clearly distinguish between clock instability and excessive (trailing edge) jitter. Remember, also, that you can set a digital scope to report the frequency of the waveform. Watch that number closely and make sure it is quite stable, otherwise you have clock instability.

What is *not* clock instability? If, for example, you have a Buck with 100% duty cycle, you may find that under transient conditions, the switch will stay completely ON for several cycles. But that is normal. Similarly, on an IC with low-side current sensing, pulses may be omitted entirely under sudden transients. That too is normal. So don't forget to *interpret* the scope waveforms you see, with due regard to the part's *architecture* and the applied conditions, not just the topology.

Once you have a stable clock, you might like to check the jitter. So how much jitter is acceptable? In an AC-DC power supply with no PFC correction, especially at low line, the input voltage ripple is quite high. As this instantaneous input voltage moves up and down, a

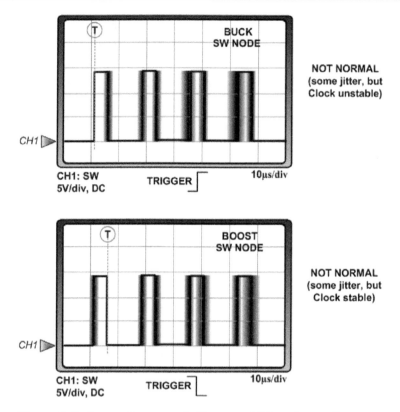

Figure 9-3 Scope Setup and Abnormal Waveforms for a Buck and a Boost

certain amount of duty cycle change is required continually to keep the output in regulation. So a certain amount of jitter will be normal. But if you have PFC, then the input to the PWM switching stage is a fairly well-regulated 385V rail. In that case, you should be very surprised if you have anything more than barely noticeable jitter. If there is excessive jitter, you may need to look very closely at the PCB layout once again. Maybe the current sense signal in your current mode control IC (the 3842) is just too noisy. Or maybe there is just not enough blanking time. However, if you try to increase the blanking time by simply increasing the time constant of the RC filter present on the current sense pin of the 3842, you should be very careful that it is not slowing down the sensed signal *too much*. Because, under a fault condition (like a sudden short on the output terminals), the speed of response of the control sections to the quickly rising current waveform may become inadequate, leading to switch failure. In other words, try always to avoid Band-Aids. They will likely leave a gaping wound elsewhere. Revisit the PCB layout and get that right first. You should also read up on Chapter 7.

Converter Instability: Staying in the Loop

Another cause of excessive jitter is loop instability. For example, you might see a fuzz around the trailing edge of modulation that is somewhat *more* periodic and refined than a random fuzz. That might actually indicate a bigger problem—that of loop instability. You should then connect your scope to the output of the power supply, put it in AC coupling mode, and zoom in really close, by adjusting the vertical divisions scale. Make sure your time division (horizontal) scale is zoomed out (or you won't see the pattern created by loop instability). If you see even a few millivolts of smoothly undulating (almost sine wave) ripple component, you probably have standard loop instability (too little phase margin). But don't forget to change your time/div setting by one click on either side, to confirm that that pattern really exists and is not just an aliasing artifact. You can also carry out a load transient test (also called a Step-Load Response Test) to look at the ringing on the output. If the output oscillates severely at each load transition, that ties in with the possibility of loop instability. In Figure 9-4, we see what 11° of phase margin can look like. In Figure 9-5, we learn how to eke out a pretty good estimate about the phase margin present just by looking at the results of the load transient test—the trick is to roughly count how many cycles it takes for the ringing to settle down. Actually, the curves in Figure 9-5 can be misleading, since they have been generated only *mathematically*, not with specific reference to a real power supply. Therefore, reliance should *not* be placed on the *amplitudes* suggested. In reality, especially when subjecting a power supply to *large-signal* load transients, say from 0A to max load, the results can change significantly from what seem to be suggested by Figure 9-5. The reason for this is that in the initial moments after the load transient, the system is dominated mainly by the amount of bulk capacitance present, *not by the loop*. The subsequent "handover" from the large-signal response to the (small-signal) loop response occurs at a rate that depends a lot on how fast the loop is. Therefore, what can really

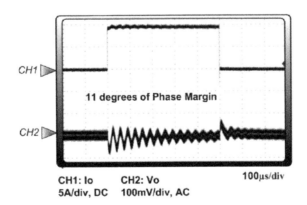

CH1: Io   CH2: Vo
5A/div, DC   100mV/div, AC   100µs/div

**Figure 9-4 Too Little Phase Margin Shows
Up in a Step-Load Response Test**

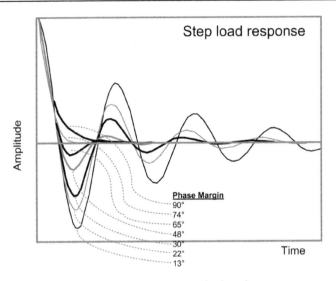

Figure 9-5 Judging Phase Margin by the Amount of Ringing in a Step-Load Response Test

happen is that if there is too much phase margin (e.g., 65° to 90°), then the handover occurs a little too late, causing the bulk capacitance to overshoot/undershoot far more severely, contrary to what Figure 9-5 suggests. So, if the final overshoot or undershoot is being ultimately determined by the bulk capacitor (as in most large-signal transients) then the effect of this slower handover will also dominate. Of course, once the handover is complete, a system with a high phase margin does tend to correct the output quickly and with less overshoot/undershoot. But what we really have in a real power supply is a battle between two somewhat opposing forces, so it is not easy to say clearly whether a large phase margin or a small phase margin will produce the least amount of overshoot/undershoot. Some experiments will need to be done if that is critical to the application. In general, it is agreed by most engineers that about 30° to 45° of phase margin gives the best compromise in terms of both the amplitude of the undershoot/overshoot and the settling time. If you want to do a formal loop measurement, follow the setup shown in Figure 9-6. Note that the wire (current loop) must be placed as shown. It should be *completely outside* any compensation components that may be present across the upper resistor of the divider (check your PCB that this indeed is possible, because this has been an all too common mistake by some engineers). Also, notice that we are using an AC probe, but its real purpose here is only to inject the signal coming from the output terminal of the analyzer into the current loop (by transformer action). Such probes have a regular BNC connector at one end and will therefore plug straight into the analyzer (or scope). Look carefully at how the two channels have to be connected. Also ensure the injected signal is not too great (to cause clipping or some strange behavior in the converter) and not too small to get drowned out by all the

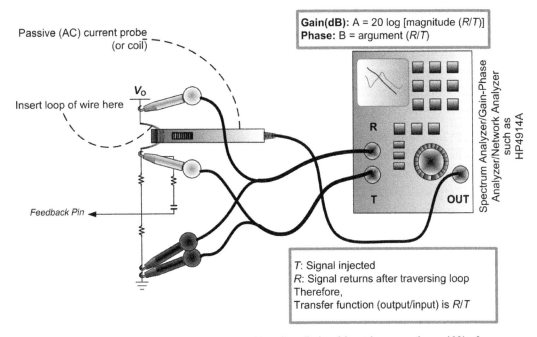

Check SW node on scope while test is in progress: The "jitter" should not be more than ~10% of the time period of the switching cycle, and not less than ~2%. Otherwise adjust Output amplitude / attenuation settings on Spectrum Analyzer.

Figure 9-6 Simple Method for a Quick Bode Plot

noise being emitted by the converter. A little playing around with the settings may be required. Note that more modern pieces of equipment allow you to preset the amplitude of the injected signal over the entire desired frequency range of interest (which is typically set from 10 or 100Hz to about 1/2 to 10 × fsw). They also call out for fancy custom-made isolation transformers, and so on, but the method shown in Figure 9-6 actually works *really well*, almost always. Note that if you find your Bode plots are too obviously dependent on the level of the injected signal, you may have a setup issue somewhere. Because they are not supposed to. The transfer function is a *ratio* between the output and the input. Having thus obtained the Bode plots, if they need fixing, you may now need to read more on the theoretical aspects of designing proper control loops. Try Chapter 6 in my book, *Switching Power Supplies A to Z.*

When working with current mode control, be aware of the problem of subharmonic instability that can occur in CCM for duty cycles greater than 50%. It has a greater likelihood of occurring at lower input voltages (in a Buck). So test at minimum input for this. It is usually recognized as an odd switching pattern consisting of one large pulse followed by a narrow pulse, repeating itself endlessly (very severe jitter but with a certain periodicity). The converter will actually seem stable to the untrained eye, especially if you don't notice the somewhat higher output voltage ripple. But the Bode plot of such a system will not be anything close to what you may have been expecting, and its step-load response will be extremely poor when operating in this mode. To suppress this instability, you will need to increase the slope compensation and/or the inductance. Or run the converter *discontinuous* (DCM or BCM).

Further, both the Boost and the Buck-Boost topologies operating in CCM (voltage-mode or current mode control) show another interesting form of instability—the RHP zero problem. This will appear as a wild, crazy, and incomprehensible jitter, with no obvious pattern to it. Stop struggling with your PCB now! The only solution here may be to roll-off the loop gain and crossover (the 0dB axis) at a much lower frequency (typically fsw/10 to fsw/20, or even less). Of course a low bandwidth is not helpful in achieving a fast transient response. But one solution, often overlooked, is to run the converter in discontinuous mode again. Yes, DCM is supposedly notorious for poor transient response, but when faced off against the RHP zero problem, it may even win! So keep an open mind.

Capturing Problems with the Single-Acquisition Mode Feature

This is probably the best feature of digital scopes when it comes to troubleshooting switching power supplies. This is where they win hands down against analog scopes. This mode can be used to capture non-repetitive events such as

a) Output voltage overshoots

b) Current overshoots

c) Power-up sequence

d) Power-down glitches

e) Input spikes at startup

f) Inrush currents

g) Soft-start

h) UVLO stuttering at startup

And so on. One of my favorite tests is the current overshoot test. I have found it extremely useful in finding subtle faults in silicon. If *anything can go wrong in the part, it is likely to express itself either as an output voltage overshoot* (but most engineers do check that well enough), *or a current overshoot*. Unfortunately, sometimes I think that either some of the engineers are getting a little too lazy to take the trouble of cutting PCB traces to insert a current probe, or there is a perennial shortage of current probes in their lab, so they wonder if it is really necessary to scour all the labs in the area for that missing current probe.

However this may have taken place, the end result is that a rigorous test of current is often overlooked.

A few years ago, our company was about to introduce a family of switcher ICs. It was to be a Buck IC with an adjustable range from 1MHz up to 3MHz. The parts had actually been declared "clean silicon" and our hallowed CEO (usually found in that fancy glass building we all called the Taj-MaHalla) was about to announce the product release from his perch somewhere in Japan, where he was on a business trip. That's when my current Boss (perennially vying to be on my list of Twenty Most Insecure People of the World) beckoned me to his cube to tell me that I was required to tie up some loose ends in the datasheet, since he had either assigned the concerned engineer to some more engaging work, or that poor guy had simply resigned (I really can't remember which). But anyway, I was used to doing housekeeping stuff like that, too (*no job should be below you in Power Conversion*), and I quickly reviewed the datasheet and made it squeaky clean for an immediate release. But having about an hour still left before going home, I thought to myself, "hey I am a Senior Engineer here after all (don't let them hoodwink me into thinking otherwise). So even though I didn't ask for it, if the part *has* come to my table, I need to show due diligence even at this late stage, and at least give it one quick look-over." So off I went to the lab, set it for the current overshoot test, not really expecting, or even mildly hoping, for any surprises. This I how I do that *current spike test.*

The Current Spike Test

I insert a current probe in series with the inductor of the Buck switcher, run it at max load and at high/low line. I set the scope on *manual trigger* for the *rising edge* of the current waveform (Channel 4 for me usually). Then I slowly move the trigger level just high enough that the scope stops triggering. I short and release the output a few times, and I expect to see the current rise smoothly and hit the current limit repetitively several times and then slide back. That's normal behavior of course. Most ICs meet this first challenge. But by doing this, I also now know exactly where the current limit of that *specific* part is. So I move the trigger level up slightly above that new level. At this point, if the part is well-designed, *under no condition should I ever be able to get the scope to trigger again.* So I start doing whatever comes to my mind, with the sole intention of trying to get the scope to trigger. I might startup into an output short, toggle the enable pin, simultaneously apply repetitive step-loading, sweep the input range, sweep the load current, change PWM to PFM modes, and so on. This may take twenty to thirty minutes or even an hour, with *nothing ever changing on the scope screen.* The longer I have to wait, the better the part (and its designer) is. At this moment, if your Boss walked in, he would simply presume you were doing nothing as usual. So it isn't a bad idea to keep a finger ready on the "Force Trigger" button. Digitals do rock sometimes!

Returning to the product release I was talking about here, it took me just five minutes to find the flaw. All I did was to set its frequency to its highest "rated" value, that is, 3MHz, and then try to get the scope to trigger. And it did, again and again whenever I shorted the output. But if I did the same at 2MHz I couldn't make the scope trigger. You have to understand the architecture of this part first. It was one of those new low-side current sense switchers. To be able to coax the highest V_{IN} to V_O conversion ratios at the highest possible frequency, you just couldn't afford the 100ns or so an IC traditionally takes to sense current in the high-side Fet. Because that would lead to an unacceptably high minimum duty cycle at high frequencies. So low-side sense switchers look (only) at the current through the *low-side Fet*, and if that samples too high, they just omit the next on-pulse altogether (until the current comes below the set threshold). One practical limitation of this architecture is that there is always a significant amount of *foldback* in the output *V-I* characteristics of the converter. So you have to set the current limit much higher than the peak current at maximum rated load. Otherwise it often latches up during startup or step-loads or overloads. What was happening in my case was that at 3MHz, whenever I would short the output, the feedback pin of the IC was getting dragged low. So the IC would respond by increasing the duty cycle to the maximum allowable, in an effort to get the output high again. And so far, this is just what should happen normally. But unfortunately, the designer had not provided *enough guaranteed minimum off-time* at 3MHz to *allow* the current sense circuitry to be able to do its job. In other words, at 3MHz, the current limit was simply nonexistent under certain conditions. What followed next were hurried parleys and consultations with the entire team. Finally the part was released, just a few days delayed, but rebranded as an adjustable 1–2MHz switcher only. I think our well-oiled CEO may have choked a little on his fried Tempura by news of the antics of this unknown troublemaker in Santa Clara. Because we had just lost our spluttering bid to reach the 3MHz mark—and thereby "show the Joneses." Though honestly, I doubt the Joneses would have been too impressed anyway, because they were already there by then. But yes, at least we could have had the thrill of going around the block saying, "Me, too."

The Hard dV/dt Test

Another test that I have found always brings out the inherent weaknesses of the part is the hard dV/dt test. Basically, I simply slam the red banana plug into the already-powered-up DC bench power supply and look for overshoots (voltage or current) in the switcher. There is a fair amount of natural input bounce created by this rather unofficial test, but that can really help aggravate/expose any startup logic issues with the IC. Of course we may later decide to specify a smooth (non-jittery) input dV/dt for the IC and just move on. My colleague used to use a mercury switch for the same purpose. That gives almost the same hard input dV/dt, but without all the bounce.

Soldering Tricks

Don't forget to have a handy Metcal soldering station right next to you. I don't know how I could ever manage without this remarkable piece of equipment. But you still have to learn some tricks on it. The packages have become so small, it is almost impossible to solder them down, even with all the help. But still, I can win almost any soldering contest hands down. Here is my way: I use a fine tweezer to place the IC on the board where it needs to be. I gently press down with the tip of the tweezer, to keep it there. Then I drop a large blob of solder over all the pins on one side of the IC. I let it cool, then I do the same on the other side. I then use solder wick to suck out all the excess solder on each side of the IC, and I actually have the most perfect joint you can imagine. In less than a minute. Also, I clean it thereafter (see the next section).

Similarly, to avoid damaging the traces when removing a damaged IC from the board, I first drop in a blob of solder on both sides of the IC. Then I use two soldering irons, one on each side, to release the IC easily, and pick it off with the help of the tips of the two irons. Once again, I use soldering wick, this time to clear up all the traces—though remembering not to use any lateral swiping (or the traces will tear away). Only press down and dab, dab, and dab.

The Fastest Troubleshooter in the World

As you know, conventional IC sockets have too much inductance to ever be used in modern switching power supplies. But, as I have said many times, it is just not enough to test one device/board and move on. We can have many boards made available to us, but they will probably last only for a few days. And besides, that also introduces too many variables in the initial stages of investigation. We would really like the flexibility of an IC socket that *works*.

In Figure 9-7 and Figure 9-8, we have the answer to all our prayers. I used it for a 600mA Buck IC project recently with excellent results. Finding a vendor for this socket is very difficult. So I decided to share the contact info of the only person I know doing this. It seems to be a one-man show, and these sockets are custom-made on demand. But they have saved me months of evaluation time. Contact "Weber" at Rainbow Labs. Address: 260 2nd Avenue, San Francisco, CA 94118, USA. Tel: 415-387-4430, Fax 415-221-3640. No, I don't know them personally, and I have no commercial stake in their operations. The information is put out only to help you, since this sort of service is very rare.

Note that these cost a couple of hundred dollars a piece usually and are mountable on standard PCB eval boards. They are actually much cheaper than the near-permanent leadless sockets some test engineers use on special boards.

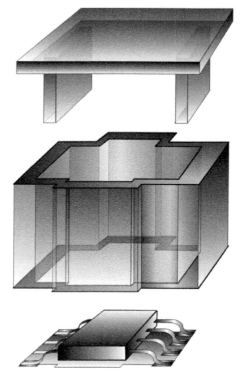

Figure 9-7 Exploded View of
Leadless Socket

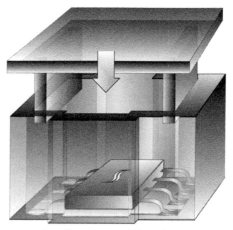

Leadless (press-fit) socket
(standard PCB mountable)

Figure 9-8 How the Leadless
Socket Works

Miscellaneous

You can get really strange results if your boards are not clean. Always take the effort to use a toothbrush with isopropyl alcohol. But note that sometimes water is also necessary. Clean the board in running water, but dry it properly before using (maybe in a warm oven). That way both organic and inorganic impurities are dissolved.

A hairdryer and a can of HFC-134a ("Freon") can help do quick temperature testing. But if you decide to put the board in a temperature-controlled oven, you should be aware that normal scope probes can melt or warp under extremes. Also choose your cables carefully (try Teflon-coated ones). Teflon cables can be very brittle, however.

Efficiency Rules

Most engineers spend a lot of their time trying to get the efficiency up to their expectations. It is often very elusive—they improve one loss term, and worsen the other. Or they end up reducing the design/derating margins, thus affecting reliability. Or they cause noticeable deterioration in some other aspect of performance. This is truly one of the most delicate balancing acts in power supply design. As indicated, we certainly can't hope to cover every aspect of this topic here. But we will try to touch on some of the most important points and common pitfalls.

However, before you start, you should have read the previous chapters and therefore actively ruled out PCB design issues, input decoupling issues, and also "junk IC" issues. You should also have asked the Twelve Questions from Figure 8-1 and assured yourself you are not obviously falling into any of the all-too-familiar traps.

Ensure the Drive Is Adequate

The first question you need to ask is, is your efficiency really bad? For example, if you have a worldwide input Flyback of around 70W, you should not be expecting much better than 70% at an input of 90VAC (for the common 5V/12V output rail combinations). For a Synchronous Buck converter, you can expect around 90% at max load, but at very light loads the efficiency will fall much lower. So first assure yourself you really have a problem. And don't forget that this measurement needs Kelvin sensing as described previously (see Chapter 2).

Now, where could the problem lie? Is it within the power supply stage itself, or is it external? If you have peeled the onion as I suggested previously, you are hopefully sure the efficiency is not being lost within the EMI filter of your AC-DC power supply. Bypass or disconnect everything external and confirm you have a problem.

The next question is, is the efficiency being lost in the *switch*? If so, there could be many reasons for that. A switch could be lossy simply because its drive is inadequate. Early self-oscillating converters (ringing choke oscillators) were extremely lossy because the drive would slowly droop to the point where it just couldn't sustain itself and then the switch would turn OFF. Modern self-oscillating converters have improved tremendously on this, and you can even find full-fledged multi-output PC power supplies that don't have a single conventional PWM control IC inside them. If you are dealing with conventional square wave converters, you need to check with a scope that you have enough drive voltage *over all parts of the switching cycle.* For example, with Buck switchers with a high-side N-Fet, it is important to check the bootstrap voltage. This is basically the supply rail for the floating driver. Connect two probes, one on the switching node ("SW") and one on the bootstrap pin (usually "Boot" or "Boost"). Don't try to connect only *one* probe between the switching node and the bootstrap pin! The voltage should therefore appear as indicated in the scope plot marked "OK" in Figure 10-1. If there are "droops," as in the plot marked "Not OK,"

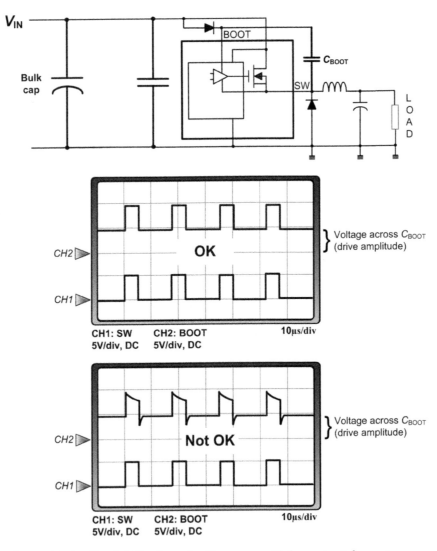

Figure 10-1 Check Whether the Bootstrap Voltage Is Adequate

you may be running out of steam towards the end of the ON time. Obviously, the worst case for this is where *D* is at its maximum. Since for all topologies, a *high D corresponds to a low* V_{IN}, you need to check the drive waveform at the lowest input voltage. The obvious fix is to increase the bootstrap capacitor, but that can also have some pitfalls as described below. It may end up even causing you to decrease the capacitor!

You should also be clear what the bootstrap circuit shown in Figure 10-1 does. When the switch turns OFF, the bootstrap capacitor gets charged up because SW has gone low. The final voltage it gets charged up to is approximately V_{IN}. Then the SW node goes high (switch ON), but the capacitor cannot lose its charge immediately. Since its lower end has gotten dragged up to approximately V_{IN}, its upper end gets hoisted to roughly $2V_{IN}$, maintaining the voltage across the capacitor at V_{IN}. See Figure 10-2 for a more detailed calculation involving the parasitic drops. You may sometimes realize you need to change over to a Schottky boot diode to maintain an adequate drive amplitude. Note that, in effect, this is just a simple *doubler* charge pump circuit. The drive amplitude that becomes available to the Gate of the Fet is approximately V_{IN} with respect to the Source. We may therefore need to look at the Fet datasheet to ensure it really does turn ON "fully" if we pull its Gate up by this amount. But we also need to protect the Fet Gate insulation, so we have to ensure that V_{IN} isn't too high. To be safe, we can always try to clamp the Gate voltage using a Gate-to-Source zener. But that is really not a good idea, at least for modern DC-DC converters, because the zener's body capacitance adds significant delay. So in many switchers designed for a *wide input*

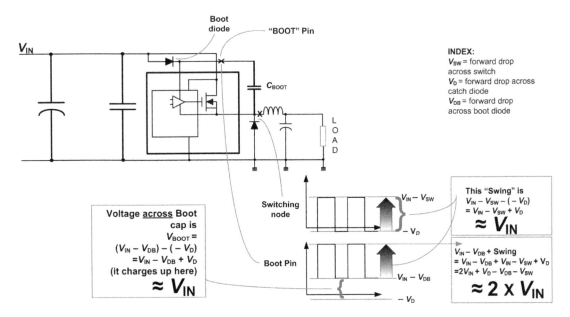

Figure 10-2 Analysis of Bootstrap Voltages in a Buck Converter

range, the bootstrap capacitor is not charged from V_{IN}, but from the ICs internally regulated rail (often called Vdd), which powers its internal control sections too. With this arrangement, the bootstrap is no longer a simple unregulated doubler—now it holds the drive voltage steady at around Vdd. But to avoid jerking the Vdd rail all over the place and upsetting the control, the bootstrap capacitor is charged from this rail by an internal current source. But that also creates a longer charging time, especially *if the bootstrap capacitor is too large*. We must therefore ensure that the OFF-time is adequate, more so with wide-input switchers, and also avoid unnecessarily large bootstrap capacitors (more important: low ESR/ESL caps).

What did I mean by "jerking?" If you try to charge an ideal capacitor with a perfect voltage source, the current demanded is theoretically infinite in the first instant. So you are literally depending on parasitics to save the show. Therefore, in the simpler bootstrap capacitor of Figure 10-1, we are demanding bursts of very high current from the input decoupling capacitor to charge up the bootstrap capacitor. And that places severe demands on the input bypassing. We can, in effect, end up injecting a great deal of noise on to the input pin. Therefore, we may need to slow this bootstrap arrangement down somewhat, too. Walking around the lab, we can often hear statements such as "lift the boot *pin* and insert a small resistor," or "lift the boot *diode* and insert a resistor." Actually, these statements are not exactly the same. Look at Figure 10-3 and see all the possible variations. The effect is different in each case. For example, if we are just haunted by jitter, but the efficiency and output noise seem OK, we would prefer to go with the middle schematic of Figure 10-3. Alternatively, if we don't think we have enough off-time to charge up the boot capacitor, but we still need to improve the output noise, we could pick the uppermost schematic. But generally, the best compromise is the lowermost schematic, "lift one end of the boot *capacitor* and insert about 10 to 20Ω." It's the *capacitor*, not the *pin*, nor the *diode*!

As you can see, we started off with a discussion on efficiency, but ended up touching noise and efficiency issues. Isn't that typical of power supply design?

Minimize Capacitive Parasitics

In Chapter 5 we discussed in great detail how parasitic trace inductances can ruin the efficiency of a typical worldwide input Flyback. Another often overlooked but significant contribution to efficiency comes from certain parasitic capacitances. These get charged up *one way* during the switch on-time and then have to discharge (or charge up the *other way*) during the switch off-time. In doing so, they invariably dump their previously stored energy as heat in the associated parasitic resistances. Some of these crucial capacitances are shown in Figure 10-4. The grayed out ones here are the parasitics—C_P is the parasitic capacitor across the primary winding, C_S across the secondary winding, and C_{OSS} across the Fet. All these are important to consider and minimize if possible. Note that sometimes, power supply manufacturers put a snubber capacitor across the switch. Its supposed purpose is to

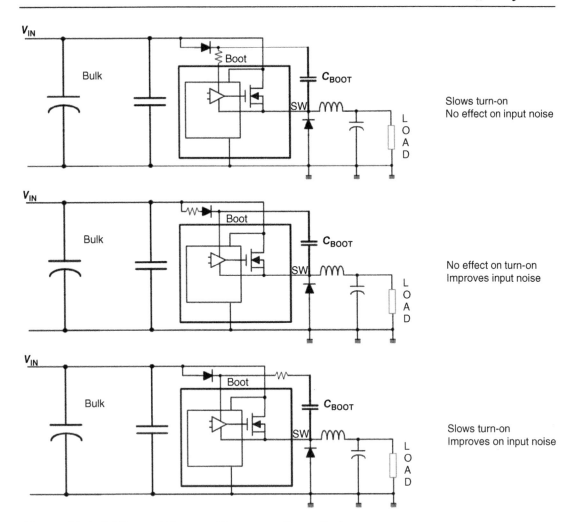

Figure 10-3 Different Ways to Connect a Boot Resistor

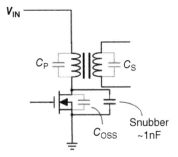

Figure 10-4 Capacitances Can Have a Key Impact on Efficiency

limit the slew rate of voltage, and thereby help reduce the switch crossover loss at turn-off, and also improve EMI. However, if a snubber is *really* required (and nowadays most people *don't* use snubbers across the switch, at least not with Fets), you should opt for a full-fledged RCD type, not a simple "capacitive snubber." The reason for this is when the switch turns ON, the capacitance dumps all its charge across it, and that too at the worst moment possible, *during* the switching transition when the voltage is still high. This increased crossover loss occurs every switching cycle. The power dissipated by it is not insignificant ($=1/2 \times C \times V^2 \times f$). The advantage of a proper RCD snubber is that a) the energy in the capacitor is dumped mainly in *R*, not in the Fet, thus saving us the problem of upgrading the heatsink. And b) it reduces the switch crossover loss during turn-off by decreasing the rising slope of voltage significantly until the current has had time to slew down to zero (its primary purpose).

Proper Design of Snubbers and Clamps

A great deal of efficiency can also be lost in a Flyback clamp. If you are using a standard zener clamp (Figure 10-5), don't miss the *correct* formula for dissipation as given below! Most people miss the very last term (to the right). For example, if we are using a 200V zener, and the V_{OR} (reflected output voltage $= V_O \times n$) is 105V, this term increases the dissipation by a factor of $200/(200 - 105) \approx 2$! The dissipation from a zener clamp can be really very high, and you can't afford to mess this calculation up.

$$P_{CLAMP} = \frac{1}{2} \times L_{LK} \times I_{PK}^2 \times f \times \frac{V_{CLAMP}}{V_{CLAMP} - V_{OR}}$$

Note that cleverly designed commercial Flybacks use an RCD clamp instead of a zener clamp. This can reduce dissipation in the clamp by around 20% at low line (where it is

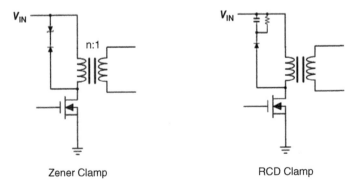

Zener Clamp   RCD Clamp

Figure 10-5 Clamps for Flyback Converters

the most significant due to the higher currents involved). But RCD clamps are considered tricky to design. Nevertheless, I remember we used them very successfully in all our Flyback designs in Singapore—and that too with only a 600V Fet (the 6N60 for a 75W supply), and there was *no additional zener clamp in parallel* to save the show if all hell broke loose (such as you will see on some eval boards from Power Integrations, for example).

The difference between an RCD snubber and an RCD clamp has nothing to do with *configuration*, that is, whether it is connected across the switch or across the transformer winding. Or even whether the resistor is across the capacitor or across the diode. All these configurations work in essentially the same way. The difference between an RCD clamp and an RCD snubber is in the *size of C*. In a snubber, the capacitor is supposed to discharge completely every cycle and then charge up again when the switch turns OFF, thereby tailoring the dV/dt that appears across the switch and reducing the *V-I* overlap, and thereby the turnoff crossover loss. In an RCD clamp, the capacitor is supposed to stay *almost* fully charged up always and simply acts as a giant reservoir that gets topped off by the energy in the leakage inductance spike. So its only purpose is to save the switch from voltage overstress. But like any inductor or capacitor involved in a repetitive switching process, *both* the capacitor in the RCD clamp and the capacitor in the RCD snubber need to reset to achieve steady state. This means they must have enough time to *discharge* the energy they picked up during the preceding turnoff transition (not necessarily all the stored energy!). Otherwise the voltage across the capacitor will staircase and could ultimately damage the switch, especially at high-line. To ensure the required amount of discharge, we need a guaranteed minimum switch on-time. And that becomes even more important for the RCD clamp/snubber, because, unlike the simple capacitive snubber discussed previously, there is now an intervening *R* in the discharge path. So we need to ensure we have *enough* minimum on-time. Everything points to a need to carefully check the operation of a snubber or clamp at highest input voltage (minimum duty cycle), under maximum load. We may have to reduce *R* to ensure capacitor reset, knowing that this will increase the dissipation in the bargain.

The key to understanding an RCD clamp is that the capacitance *C* doesn't really enter the picture, at least not in a first-order calculation (unless it is very small). Because eventually, it tends to automatically stabilize at the same *average* level, and that depends on the value of *R*. In any case, the capacitance needs to be quite high (around 22nF), since under a sudden short-circuit, the capacitor will start charging up quite quickly, and we don't want to exceed the voltage rating of the switch before the current limit chimes in to limit this abnormal condition. Once *C* has been picked (hopefully high enough), *R* should be set/ adjusted at high line, under normal operating conditions and max load, such that there is still about 50V margin remaining. In other words, if we are using a 600V Fet, then at 270VAC and max load, *R* should be set so that V_{DS} does not exceed about 550V. That usually leaves enough margin for any properly designed current limit circuit to act *just in time* (aided of course by a properly designed line feedforward), during output shorts or overloads.

Since the design of an RCD clamp is critical and tricky, a formal design procedure is now provided, with reference to a useful nomogram provided in Figure 10-6. Note that these particular curves are set up for a 700V Fet and a (common) V_{OR} (reflected output voltage) of 105V. The basic design procedure is very simple.

1. Measure the in-circuit primary-side leakage inductance L_{LK} in Henries.

2. Measure the peak of the Drain current waveform I_{LIM} at rated power at 90VAC.

3. Calculate $E_{LK} = \frac{1}{2} \times L_{LK} \times I_{LIM}^2$ µJ.

4. Tentatively select one of the three clamp capacitor values 4.7nF, 10nF, or 22nF (prefer the 22nF usually).

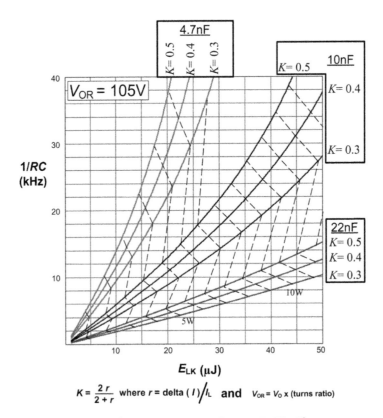

Figure 10-6 Design Nomogram for an RCD Clamp

5. Draw a vertical line from the calculated E_{LK} to intersect with the solid line curve corresponding to the selected C and K.

6. Interpolate between the dotted line curves to estimate the clamp dissipation.

7. If the dissipation is considered acceptable, the *y*-coordinate provides $1/RC$. Calculate R.

It is important to get this calculation right if you are seriously worried about the efficiency of your Flyback. But don't forget to do a final bench verification—short and overload the output at high line, and capture the peak voltage stress on the Fet.

Also, if your clamp is too hot, check that the leakage inductance of the transformer is better than about 1 to 2% of the primary inductance. If not, you need to first reduce the leakage inductance. That calls for improving the coupling between the Primary and (main) Secondary windings. Move them up closer toward the safety insulation barrier. Interleave/ sandwich for anything greater than 40W output power—and that usually calls for a split primary (both halves in series), placed on either side of the secondary winding. Check that you don't have noise screens at the safety interface. These invariably increase leakage by pushing the primary and secondary windings further apart. Also ensure you have only three layers of polyester tape insulation at the safety interface, not more. You may like to reduce the thickness of the tape from 2 mil to 1 mil or even 1/2 mil. All these thicknesses are actually allowed by safety regulations, provided they meet the stipulated dielectric withstand voltage.

If you can't reduce the dissipation in your clamp by any other means, try to exploit the fact that increasing $V_{CLAMP} - V_{OR}$ will reduce the dissipation. So you can try to increase V_{CLAMP} or reduce V_{OR}. To increase V_{CLAMP}, try increasing the zener voltage of your zener clamp, or the R of your RCD clamp, ensuring that in the process you are not in danger of ever exceeding the voltage rating of the switch. To decrease V_{OR}, you need to reduce the turns ratio (since $V_{OR} = n \times V_O$). Note that the duty cycle of a Flyback is

$$D = \frac{V_{OR}}{V_{IN} + V_{OR}}$$

In other words, the transformer-based Flyback behaves just like an inductor-based Buck-Boost, with the difference that the "output voltage" is V_{OR}, not V_O. So decreasing V_{OR} calls for a decrease in D. However, the same input power still has to be drawn from the switch. So if the width of its waveform decreases, the height of the waveform must increase. Which means that the inductor current must also increase. So, decreasing V_{OR} could also end up decreasing efficiency. That is why for most universal-input Flybacks, the best V_{OR} compromise is about 90V to 105V.

There is something puzzling about the statements above in case you haven't noticed! How are we concluding that a decrease in D causes an increase in the inductor current? So far we have been led to believe that in a Buck-Boost or Boost topology, the inductor current equals $I_O/(1 - D)$, which implies that the inductor current goes up as D increases, not decreases! So, is that not true for a Flyback? Actually it is, because $I_L = I_{OR}/(1 - D)$. But what has happened here is that by *changing the turns ratio*, we changed I_{OR}, too, because $I_{OR} = I_O/n$. So by decreasing the turns ratio n, we have actually increased I_{OR}, and therefore also I_L. But, *if we keep the turns ratio fixed*, the Flyback certainly follows the known behavior of the Buck-Boost and Boost with respect to changes in D.

The last resort for reducing the clamp losses is to reduce the switching frequency, since the clamp loss term is purely a switching loss term and is therefore proportional to the switching frequency.

Varying the Frequency

Looking hard at the switch and diode now, the question always is, is the efficiency loss due to excessive crossover (or switching) losses or is it due to conduction loss? Note that in principle, though the diode has conduction losses, it has no (V-I) crossover loss, because the transition is essentially driven by the switch and so there is no remaining V-I overlap across the diode. We also know that conduction losses do not depend on frequency. But by definition, switching (or crossover) losses do—they are supposed to be *proportional* to frequency. So let us assume our control IC offers us the option of varying the switching frequency. If we find the efficiency falls a little too steeply as we increase the frequency, that could indicate excessive switching losses. Of course there is some judgment involved concerning what exactly constitutes "too steeply." We also want to be able to *isolate the crossover losses of the switch* to be able to study them more closely. For example, if we have a typical 5V/12V multiple output AC-DC power supply, we may like to disconnect the 12V rail for this investigation. That is because the 5V diode is usually a Schottky diode in these applications, and we know that has almost no *reverse-recovery current* switching loss term, whereas a regular ultra-fast diode does. Yes, we could also have other significant frequency-dependent terms, as in the transformer, so we may like to monitor it to see whether it is getting too hot. Also the clamp as mentioned above. An efficiency investigation is always slow and painful, and usually involves incremental and studied improvements at different points inside the converter, rather than one giant leap.

The Time-sharing Principle

Now, we have an important principle to understand, what I call the *principle of time sharing*. To understand it, we will need to start by applying it only to DC-DC converters at first. Later, we will learn it applies well to any Buck converter, but not to the other topologies. We will *be keeping the output voltage fixed* in the following discussion, because otherwise we will be guilty of trying to compare apples with oranges. For example, a 5V/1A converter with a total loss of 0.5W has an efficiency of $P_O/P_{IN} = 5/5.5 = 91\%$. But a 10V/1A converter with a *higher* loss of, say, 0.7W, has in fact a *better* efficiency ($10/10.7 = 93.5\%$), despite the load current being unchanged. It can all get very confusing.

The time-sharing principle tells us basically how the *total conduction losses* (in the switch and diode combined) behave with respect to input voltage, *depending on the ratio V_{SW}/V_D*, where V_{SW} is the drop across the switch, and V_D is the drop across the diode. If we decrease D, and if $V_{SW} > V_D$ (usually the case), the current will spend less time going through the higher drop, and so the total conduction loss will decrease and efficiency will increase. Note that *for all topologies, a high D corresponds to a low input and vice versa*. So that implies that at high input voltages, we expect the efficiency of any typical Non-Synchronous converter to increase. Note that if the switch happens to be a Fet instead of a BJT, its drop V_{SW} is in effect a function of the current through it ($V_{SW} = I_L \times$ Rds, where I_L is the average inductor current). So sooner or later, at high enough load currents, we will get the same result, $V_{SW} > V_D$, and the efficiency will again increase as the input voltage is raised.

Coming to Synchronous converters, we have two Fets sharing time over each switching cycle. One of these Fets is placed where the diode would have been in the corresponding Non-Synchronous topology. This is therefore called the "Synchronous Fet" whereas the other becomes the "control Fet." In principle, the Synchronous Fet (like the diode) sees no crossover losses, because its voltage and current waveforms do not overlap. So its switching speed is not usually of primary concern—rather it is optimized for lowest conduction losses, and therefore is typically a low-Rds Fet. The drop across it (i.e., in effect V_D) is therefore comparatively low. The control Fet on the other hand is optimized for low crossover losses (higher speed) and therefore usually has a somewhat higher Rds. In effect, it has a high V_{SW}. In other words, the situation becomes similar to that in a Non-Synchronous topology ($V_{SW} > V_D$), and so the efficiency again improves as input voltage increases. The only difference is that *in effect the time-sharing principle becomes a comparison of the Rds of the two Fets.*

What happens if $V_{SW} < V_D$? In fact that is the situation in most commercial Flybacks. But note that to do a proper comparison, you have to *reflect the diode drop* to the primary side. And for that we have to multiply the diode drop by the turns ratio (see the "equivalent Buck-Boost models of a Flyback" section in my book, *Switching Power Supply Design & Optimization*). So, for example, if the turns ratio is 20 and the diode drop is 0.6V, the effective V_D we need to compare with V_{SW} for our time-sharing analysis is $0.6 \times 20 = 12V$. And that is usually greater than the (average) drop across the switch. Therefore, we tend to say that in a Flyback, decreasing D (increasing input) will worsen the total conduction loss and decrease the efficiency. But of course that never happens, because as we increase the input (decreasing D), the inductor current falls off dramatically because $I_L = I_O/(1 - D)$. And that reduces the total conduction loss significantly, irrespective of what the time-sharing principle seems to suggest. So efficiency tends to always improve as we increase the input voltage, irrespective of whether $V_{SW} > V_D$ or $V_{SW} < V_D$. In other words, the time-sharing principle is a nice tool for understanding Buck converters, Synchronous or otherwise, but not necessarily the other topologies.

In all cases, in any topology, irrespective of V_{SW} and V_D, as we increase the input voltage sufficiently, the switching losses will ultimately start predominating and the efficiency will roll off.

But we may succeed in pushing this roll-off point further and further away by reducing switch transition times.

Understanding the Shape of Buck Efficiency Curves

By the time-sharing principle, we see that in a Buck converter if V_{SW} is close to V_D, the conduction losses do not change with duty cycle or input voltage. But the switching losses progressively increase, and so the efficiency falls off smoothly (almost linearly) with increasing input. See Figure 10-7 for the curve marked $V_{SW} = V_D$. An example of this is the

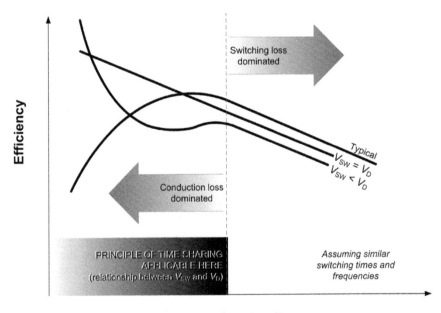

Figure 10-7 Possible Variations of Efficiency Curves for Buck Converters

published efficiency curve of the LTC3835 (available at www.linear.com). This is a Buck controller, but the *same* Fet has been selected for both upper and lower positions on the typical applications board. Also see another example, the efficiency curves of the MAX8506 integrated Synchronous Buck from Maxim (at www.maxim-ic.com), which has two Fets of almost the *same* Rds.

We also have some industry cases where $V_{SW} < V_D$. An example of this is the LTC1877. Its efficiency versus input voltage curve resembles the odd curve marked $V_{SW} < V_D$ in Figure 10-7. That curve is clear evidence of a strange switch/diode drop ratio.

Efficiency of Universal Input Flybacks

Typically, most worldwide input Flybacks exhibit their lowest efficiency at low line (90VAC). That is because in the Flyback topology, currents are highest at lowest input voltages. So the conduction loss term predominates at low inputs. But note that crossover loss is actually a cross-product of *both* voltage and current. So the crossover loss is not necessarily insignificant at low line. However, in general, switching losses tend to predominate only at high input voltages. So we get the typical efficiency curve seen in the upper half of Figure 10-8. Note that it tends to max out somewhere in the middle, between 90VAC and 270VAC. But despite that, *usually, the efficiency at 270VAC is better than the efficiency at 90VAC.*

The above-described behavior is merely typical. For example, there is a relatively new breed of ultra-low Rds Fets such as the *CoolMos* from Infineon AG. They do represent an exciting milestone in technology—I remember even the International Rectifier sales rep was running scared from them a few years ago in a presentation to us. But one of my colleagues reported at around the same time that the Flyback converter he had just built using this device had much-improved efficiency at low line, but that the efficiency at high line was now *worse* than the efficiency at low line (see the upper half of Figure 10-8). That was the first time we saw something like that! It seemed obvious that in the brute-force attempt to reduce Rds, the switching speed of these devices had been compromised noticeably. But also remember, this was the situation quite a few years ago. This was just the first-generation CoolMos (called S5). These devices are considerably better today.

So far we have been looking very closely at the variation of efficiency with line. We can also ask, how does the efficiency vary with respect to *load current*? At very light loads, the system enters discontinuous conduction mode (DCM). In this mode, the edge of inductor current the switch turns ON into is zero, so that gives *zero turn-on* crossover loss. But despite that narrow advantage, since the shape of the switch current becomes very peaky, its RMS value is relatively higher, and so is the switch conduction loss (of course everything being relative to the useful power delivered). That is one of the reasons why at light loads the efficiency always starts falling. See the lower half of Figure 10-8. Another reason is that

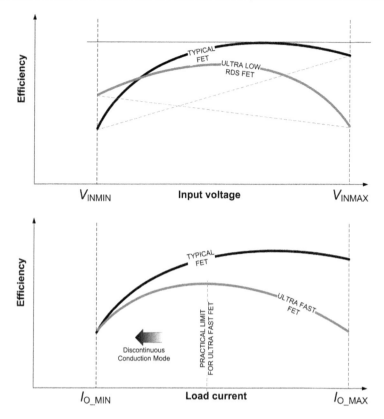

Figure 10-8 Typical Flyback Efficiency Curves

any circuit has a supply current component that is relatively fixed, such as the current required to bias its bandgap reference, its comparator, its error amplifier, and so on. There is also probably a clock running full time. There could also be external components, such as the voltage divider, that continue to draw the same amount of current irrespective of loading. And all of these become a bigger proportion of the power being transferred. So the efficiency starts to plummet at light loads.

One thing is clear—*usually, the efficiency peaks very close to the maximum rated load current.* That is actually considered natural in most mechanical systems, too—any normally designed car engine, for example (without overdesign *or* underdesign) is the most efficient when operated at its maximum capacity. But if you look at the efficiency curves of some high-voltage switchers such as the Topswitch (from Power Integrations at www.powerint. com), you will see a rather atypical efficiency curve. It looks a lot like the curve labeled ultra-fast in the lower half of Figure 10-8. Yes these devices are probably considered ultra-fast (at least by their manufacturers), but I personally consider them as being virtually the opposite of CoolMos, that is, ultra-*high* Rds! Their efficiency is clearly dominated very strongly by conduction losses, and that is why the efficiency falls almost constantly with increasing load current. Note that these devices are rather astutely rated, not in terms of the maximum *practical* power they can deliver, but by the very simple *electrical* criterion, "at what load does the peak current hit the current limit?" The current limit itself is, incidentally, positioned at the very edge of where the device can no longer function as a switch (barely even as a semiconductor). In fact at that edge, the Drain-to-Source voltage drop is a stupendous 18V for the entire family (at 100°C). Therefore, in such cases, rather than struggling endlessly with efficiency, and bigger and bigger heatsinks, recognize the limitations of your device and simply move on to a device with a higher *declared* power rating than you need. In other words, ignore the hyped maximum rated power, and *look at the actual Rds*. That makes far more sense to an engineer anyway.

We learn some fairly basic lessons to help us in our troubleshooting efforts from all the different efficiency curves presented above. So, if we now see the efficiency plummet at high line, we probably have a switching loss problem somewhere. If the efficiency plummets at high load currents, we probably have a high conduction loss problem. Admittedly, we can always try searching our inventory endlessly in search of the ideal Fet. But like the ideal mate before it, there is none! If we pick a very low Rds Fet, we will almost certainly suffer higher switching losses, and vice versa. The latest generation CoolMos claims you can finally have your cake and eat it too—the best Rds $\times$ Q_G of the industry. Therefore I found it surprising that all their low-power demo Flyback boards seemed to have been designed *only for DCM*. I started thinking that maybe their Apps guys know something I don't.

Estimating the Ratio of Conduction Loss to Switching Loss

Having understood how any measured efficiency curve reflects on the conduction losses taking place in the power stage vis-à-vis its Rds/V_{SW} and diode drops, we can now actually start putting numbers to all of this. *If we pick any two points on the efficiency curve for a given output voltage, we can actually find out a surprisingly lot of information on what is happening there*. We can also use that information to generate efficiency points for any other output voltage. In the process, we are also going to find out how much of the total switch plus diode losses are conduction losses, and how much are crossover losses.

Let us pick four points along a set of typical efficiency curves as shown in Figure 10-9. We analyze this in Figure 10-10, and the steps should be fairly obvious. Basically, we are writing the loss at each point as the sum of the switch conduction loss, the diode conduction loss, and a generic switching loss (crossover) term. We thus arrive at the general solution to the equations. We then take the published efficiency curves for the 2593HV (see

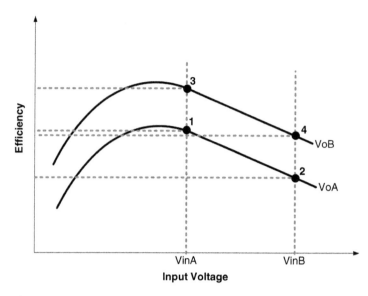

Figure 10-9 Analysis of Typical Efficiency Curves of a Buck (Part 1)

We can write general equation:

$$\textbf{Loss}_1 = I_O \times V_{SW} \times D_1 + I_O \times V_D \times (1 - D_1) + k \times I_O \times V_{INA}$$
$$\textbf{Loss}_2 = I_O \times V_{SW} \times D_2 + I_O \times V_D \times (1 - D_2) + k \times I_O \times V_{INB}$$
$$\textbf{Loss}_3 = I_O \times V_{SW} \times D_3 + I_O \times V_D \times (1 - D_3) + k \times I_O \times V_{INA}$$
$$\textbf{Loss}_4 = I_O \times V_{SW} \times D_4 + I_O \times V_D \times (1 - D_4) + k \times I_O \times V_{INB}$$

Usually,
$k = \text{tcross} \times f$ where 'tcross' is the crossover time (considered unknown here)

where,

$$D_1 = \frac{V_{OA}}{V_{INA}}, \quad D_2 = \frac{V_{OA}}{V_{INB}}, \quad D_3 = \frac{V_{OB}}{V_{INA}}, \quad D_4 = \frac{V_{OB}}{V_{INB}}$$

Call "Loss/I_O" as "loss" i.e. loss per unit load current, and "$V_{SW} - V_D$" as "v":

$$\textbf{loss}_1 = V_{SW} \times D_1 + V_D \times (1 - D_1) + k \times V_{INA}$$

$$\textbf{loss}_1 = v \times D_1 + V_D + k \times V_{INA}$$

$$\textbf{loss}_1 = v \times \frac{V_{OA}}{V_{INA}} + V_D + k \times V_{INA}$$

$$\textbf{loss}_1 \times V_{INA} = v \times V_{OA} + V_D \times V_{INA} + k \times V_{INA}^2$$

By definition:

$$Loss = \frac{1 - \eta}{\eta} \times P_O$$

$$Loss = \frac{1 - \eta}{\eta} \times V_O \times I_O$$

$$loss = \frac{1 - \eta}{\eta} \times V_O$$

where 'η' is the Efficiency

Similarly for another input voltage point (on the same output voltage curve):

$$\textbf{loss}_2 \times V_{INB} = v \times V_{OA} + V_D \times V_{INB} + k \times V_{INB}^2$$

We have two equations for "loss$_1$" and "loss$_2$" and two unknowns, v and k. Solving, we get

$$k = \frac{(\textbf{loss}_1 \times V_{INA} - \textbf{loss}_2 \times V_{INB}) - V_D (V_{INA} - V_{INB})}{V_{INA}^2 - V_{INB}^2} \qquad v = \frac{\textbf{loss}_1 \times V_{INA} - V_D \times V_{INA} - k \times V_{INA}^2}{V_{OA}}$$

Note that $V_{SW} = v + V_D$

Knowing k, and v, we can predict efficiency at any other point (even for different output voltages)

See worked example in Mathcad Box on pages 224 to 225

Figure 10-10 Analysis of Typical Efficiency Curves of a Buck (Part 2)

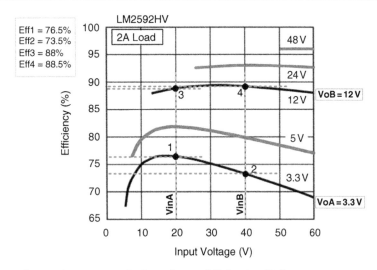

Figure 10-11 Analyzing the Published Efficiency Curves of the 2592HV

Figure 10-11), and apply the analysis. We can use the Mathcad Box that follows on the next page (or tedious hand calculations) to predict the efficiencies of other points, and also predict the percentage of switching losses. As we can see, we get extremely good agreement with measured results. In fact, if we know any two points of an efficiency curve, we can predict the entire efficiency behavior of the part—for any output voltage, any input voltage, and even any load current. We also discover the effective switch drop and the transition time. Try it! It can save a lot of lab time (though I don't recommend this during evaluation).

Conversely, if the predictions are not matching the measured results, the following must be considered:

a)   Is the switcher heating up too much and affecting switch drop?

b)   Is there loop instability?

c)   Is the switch drive really adequate?

d)   Is the switch drop (or Rds) a steep function of voltage?

e)   Are other component losses playing a more significant part?

f)   Are the magnetics or the capacitors getting too hot?

g)   Is the IC entering some different mode of operation (like DCM or PFM)? And so on! Almost everything else (sorry)!

$$I_O := 2$$

$$V_D := 0.5$$

$$V_{INA} := 20$$

$$V_{INB} := 40$$

$$V_{OA} := 3.3$$

$$V_{OB} := 12$$

Measured Points:

$$Eff1 := 0.765$$

$$Eff2 := 0.735$$

$$Eff3 := 0.88$$

$$Eff4 := 0.885$$

Since "loss" is defined as Loss/I_o,

$$loss_1 := \frac{1 - Eff1}{Eff1} \cdot V_O A$$

$$loss_2 := \frac{1 - Eff2}{Eff2} \cdot V_O A$$

$$loss_3 := \frac{1 - Eff3}{Eff3} \cdot V_O B$$

$$loss_4 := \frac{1 - Eff4}{Eff4} \cdot V_O B$$

$$k := \frac{(loss_1 \cdot V_{INA} - loss_2 \cdot V_{INB}) - V_D \cdot (V_{INA} - V_{INB})}{V_{INA}^2 - V_{INB}^2}$$

$$V_{SW} := \frac{loss_1 \cdot V_{INA} - V_D \cdot V_{INA} - k \cdot V_{INA}^2}{V_{OA}} + V_D$$

$$V_{SW} = 1.864 \ (\text{SWITCH DROP})$$

Use this to predict any other points

$$D3 := \frac{V_{OB}}{V_{INA}}$$

$$D4 := \frac{V_{OB}}{V_{INB}}$$

$$loss_{3\_CALCULATED} := I_O \cdot V_{SW} \cdot D3 + I_O \cdot V_D \cdot (1 - D3) + k \cdot I_O \cdot V_{INA}$$

$$loss_{4\_CALCULATED} := I_O \cdot V_{SW} \cdot D4 + I_O \cdot V_D \cdot (1 - D4) + k \cdot I_O \cdot V_{INB}$$

$$\text{Eff}_{3\_\text{CALCULATED}} := \frac{V_{\text{O}} \text{B} \cdot I_{\text{O}}}{\text{loss}_{3\_\text{CALCULATED}} + V_{\text{O}} \text{B} \cdot I_{\text{O}}}$$

$$\text{Eff}_{4\_\text{CALCULATED}} := \frac{V_{\text{O}} \text{B} \cdot I_{\text{O}}}{\text{loss}_{4\_\text{CALCULATED}} + V_{\text{O}} \text{B} \cdot I_{\text{O}}}$$

$$\text{Eff}_{3\_\text{CALCULATED}} = 0.882$$

$$\text{Eff}3 = 0.88 \ \text{MEASURED (COMPARE)}$$

$$\text{Eff}_{4\_\text{CALCULATED}} = 0.89$$

$$\text{Eff}4 = 0.885 \ \text{MEASURED (COMPARE)}$$

Percentage of Switching Loss

For point 3

$$\frac{\text{k} \cdot I_{\text{O}} \cdot V_{\text{INA}}}{\text{loss}_{3\_\text{CALCULATED}}} = 0.18$$

For point 4

$$\frac{\text{k} \cdot I_{\text{O}} \cdot V_{\text{INB}}}{\text{loss}_{4\_\text{CALCULATED}}} = 0.388$$

Losses in Input/Output Capacitors and Magnetics

Let us do a brief survey of how the losses in the related components play out.

First a note of caution. Don't always rely on intuition to get you by. For example, as you increase the input voltage in a Buck (i.e., decrease D), the peak current increases and the switch waveform becomes more peaky, as is obvious from Figure 10-12 (r is defined as $\Delta I / I_{\text{L}}$). We intuitively expect this to increase the RMS current in the switch, and also the RMS current in the input and output capacitors. But that does not really happen as you can see in Figure 10-13, with reference to its index in Table 10-1. For example, curve number 7, representing the RMS current in the switch actually falls as D decreases. That is because, even though the waveform became more peaky, the decreasing width of the switch current waveform more than made up for it, and therefore the RMS switch current decreases with increasing input. In fact, look also at the RMS switch currents for the remaining topologies (curve number 2). We can therefore decisively state that for *all topologies*, increasing the input voltage decreases the switch conduction loss (period)! What impact that has on the overall efficiency depends on several factors. One is the time-sharing principle explained earlier, which asks, "though you have decreased the switch conduction loss, are you sure the increased diode conduction loss has not more than offset that advantage?" But besides the diode, we could be seeing severe losses in other components, too, that we need to check out.

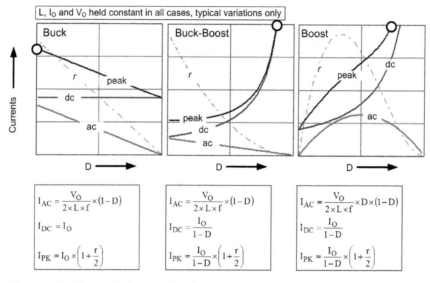

$$I_{AC} = \frac{V_O}{2 \times L \times f} \times (1 - D)$$

$$I_{DC} = I_O$$

$$I_{PK} = I_O \times \left(1 + \frac{r}{2}\right)$$

$$I_{AC} = \frac{V_O}{2 \times L \times f} \times (1 - D)$$

$$I_{DC} = \frac{I_O}{1 - D}$$

$$I_{PK} = \frac{I_O}{1 - D} \times \left(1 + \frac{r}{2}\right)$$

$$I_{AC} = \frac{V_O}{2 \times L \times f} \times D \times (1 - D)$$

$$I_{DC} = \frac{I_O}{1 - D}$$

$$I_{PK} = \frac{I_O}{1 - D} \times \left(1 + \frac{r}{2}\right)$$

Figure 10-12 Variations of Inductor Current Components for DC-DC Converters

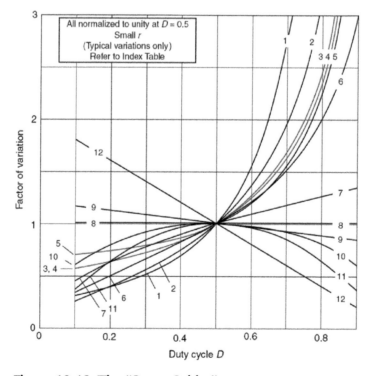

Figure 10-13 The "Stress Spider"

Table 10-1 Index Table for the "Stress Spider"

Parameters	Buck	Boost	Buck-Boost
Inductor Current Swing $\Delta I(2 \times I_{AC})$	V_{IN_MAX} 12	V_{IN_50} 11	V_{IN_MAX} 12
Core Loss	V_{IN_MAX}	V_{IN_50}	V_{IN_MAX}
Inductor Energy/Core Saturation	V_{IN_MAX}/V_{IN} 8	V_{IN_MIN} 1	V_{IN_MIN} 1
Average Current in Inductor	V_{IN} 8	V_{IN_MIN} 3	V_{IN_MIN} 3
RMS Current in Inductor	V_{IN_MAX}/V_{IN} 8	V_{IN_MIN} 3	V_{IN_MIN} 3
Copper Loss/Temperature of Inductor	V_{IN_MAX}/V_{IN}	V_{IN_MIN}	V_{IN_MIN}
RMS Current in Input Capacitor	V_{IN_50} 10	V_{IN_50} 11	V_{IN_MIN} 6
Input Voltage Ripple	V_{IN_MAX}/V_{IN} 8	V_{IN_MAX} 12	V_{IN_MIN} 3
RMS Current in Output Capacitor	V_{IN_MAX} 12	V_{IN_MIN} 6	V_{IN_MIN} 6
Output Voltage Ripple	V_{IN_MAX} 12	V_{IN_MIN} 3	V_{IN_MIN} 3
RMS Current in Switch	V_{IN_MIN} 7	V_{IN_MIN} 2	V_{IN_MIN} 2
Average Current in Switch	V_{IN_MIN}	V_{IN_MIN}	V_{IN_MIN}
Peak Current in Switch/Diode/Inductor	V_{IN_MAX} 9	V_{IN_MIN} 4	V_{IN_MIN} 5
Average Current in Diode	V_{IN_MAX} 12	V_{IN} 8	V_{IN} 8
Temperature of Diode	V_{IN_MAX} 12	V_{IN} 8	V_{IN} 8
Worst case Efficiency (typical)	V_{IN_MAX}	V_{IN_MIN}	V_{IN_MIN}

Numbers in the columns refer to corresponding numbered curves in the "Stress Spider"
V_{IN} simply means any input voltage is appropriate
V_{IN_50} is input voltage at which $D = 0.5$

For the Buck, we see that the input capacitor RMS actually maxes out at $D = 50\%$, whereas the output capacitor RMS current (curve number 12) increases dramatically at low D (high input). Does that really mean that we have to worry about the dissipation in the output capacitor? Think about it. The output capacitor in a Buck is barely responsible for any of its losses, since it sees only the smoothened (undulating) inductor current. So yes, as a *percentage* it does goes up as D decreases, but its *absolute value* is still usually negligible. The only loss component that really goes up as we increase the input voltage is the core loss, since that depends on the *swing* of current ΔI, which for a Buck or a Buck-Boost goes up strongly as D decreases. See the curve marked 12 in Figure 10-13. Surprisingly for some, the core loss of a Boost is at a maximum when $D = 0.5$ (or the point closest to it for our input range). See the curve marked 11 in Figure 10-13. This is also confirmed by Figure 10-12. So if your efficiency at high line is falling much too steeply in your Buck or Buck-Boost converter, check if your inductor is getting too hot. You may like to increase the inductance somewhat, as this will decrease ΔI, and thus the core loss. If copper loss is the culprit, decreasing the number of turns will help.

CHAPTER **11**

Magnetics, EMI, and Noise

The Wish List

Several years ago, a senior engineer from a well-known U.S.-based OEM visited us in Singapore to "try us out." What his employer shared with the company we were already doing business with was that both were massive corporations that had had their humble (but spectacular) beginnings in *garages*. But this engineer also produced a strange wish list— *make me the quietest 65W Flyback ever.*

How quiet? A standard noise and ripple measurement setup is shown in Figure 11-1. Between the power supply and the load (in this case a board with appropriate high-wattage resistors) lies a fairly long cable harness. Since the power supply is open-frame, the cables leaving the power supply always manage to pick up a great deal of the noise being emitted from it (via radiation or conduction). And since the scope probe is to be hooked up at the *load end*, a small 0.1μF disc ceramic capacitor is (*almost*) always allowed to be present at the point of measurement (as it would be in real-world systems, too).

I said "almost," because the visiting engineer had just thrown us an interesting challenge (in between all the caviar my Boss was busy stuffing him with). He wanted the "usual" maximum of ±50mV noise and ripple on the 5V output, measured with a full 100MHz bandwidth analog scope (with automatic cursors that include the noise you can't see on the screen), but *without the 0.1μF capacitor.* I wondered what he had been drinking.

As you guessed, I was the struggling power supply design engineer assigned to satisfy him. Wondering if I would lose my job over this, I finally got lucky after about a month. Then, donning some freshly discovered robes, I delivered him his wish list. He wasn't expecting it for sure. After sobering up he finally emailed my Boss, "How does your company make such incredibly quiet Flybacks after all?" Of course he was just being difficult, because *that* stringent a spec was never really required for any equipment he would ever build. And as predicted, it wasn't ever built! At least not in *exactly* the way I had created it. However, my company did acknowledge that I had somehow managed to create a very useful bag of tricks for them to use. Some of those learnings did find their way into the very next model my company made for this OEM (using the spec that was *really needed*). What they didn't implement was my admittedly exotic transformer design

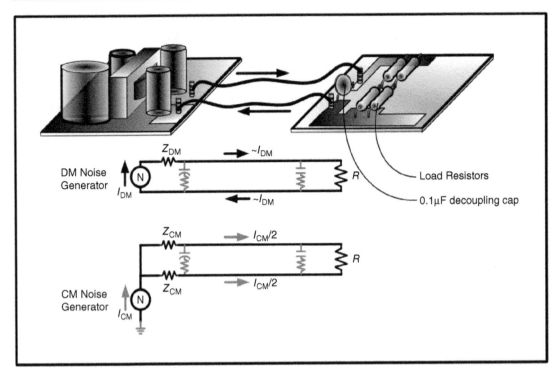

Figure 11-1 One Possible Way of Conducting a Noise and Ripple Measurement for an AC-DC Power Supply

(they considered it not "productionizable"). Or my RCD clamp design ("not necessary"). But not being one to give up so easily, I will talk about *both* of these here, though just a little later. At this point I only want to draw your attention to the last nagging problem I faced, because therein lies a fundamental lesson about EMI for all of us.

The problem after the 20th day of trial and error was that I could easily meet the requirement on the setup shown in Figure 11-1 (without the disc ceramic capacitor as demanded), but I couldn't still meet it on the setup shown in Figure 11-2 (still with no 0.1µF capacitor). What is the difference? In the latter figure, the only change is the thick metal sheet below the power supply and its load. The purpose of that was to simulate what would finally be the customer arrangement, with both the power supply and the system board present inside a large metal enclosure (the chassis). Of course I still question the validity of connecting *both* the power supply (its secondary ground) *and* the system board ground plane to the metal plate by metal standoffs as shown in the figure. The ohmic drops across the output cables can create a big ground loop with circulating currents through the chassis which causes them to radiate. Nowadays, it is more common to first bring out all the output cables of the power supply to the system board,

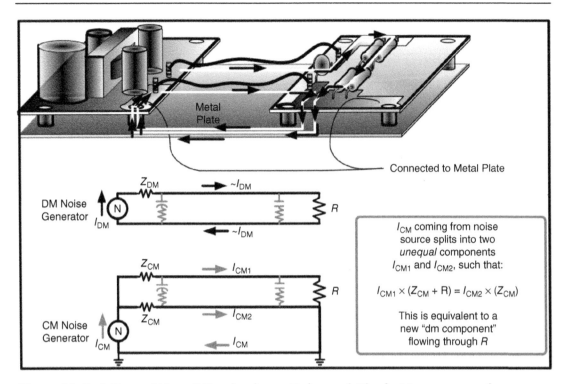

DM Noise Generator
Z_{DM} ~I_{DM} R ~I_{DM} I_{DM}

CM Noise Generator
Z_{CM} I_{CM1} R Z_{CM} I_{CM2} I_{CM}

Connected to Metal Plate

Metal Plate

I_{CM} coming from noise source splits into two *unequal* components I_{CM1} and I_{CM2}, such that:

$$I_{CM1} \times (Z_{CM} + R) = I_{CM2} \times (Z_{CM})$$

This is equivalent to a new "dm component" flowing through R

Figure 11-2 A Better Way of Conducting a Noise and Ripple Measurement for an AC-DC Power Supply (But Not the Best)

and to connect to chassis ground *only* at that point. This is shown in Figure 11-3 (more on that later, too). But Figure 11-2 is how we were used to performing a Noise and Ripple measurement in Singapore those days (with the 0.1μF capacitor present, until this funny guy came along). In fact Figure 11-2 is exactly what we also used for the standard CISPR22 conducted and radiated emissions tests. Though I remember that arrangement did come to bite our first OEM eventually. Just a few weeks short of a major product release, they were getting noticeable mysterious dark bars rolling across their computer screens. The proprietor of our Singapore company had just returned from their huge production facility (also in Singapore) and reported that almost all their engineers were literally sprawled across the shop floor trying to fix the grounding arrangement of their computer chassis. New slots were being punched, holes drilled, almost without a clue, with only the final results expected to shed light on the way out of the tunnel. The proprietor was chuckling, "It looks like they have gone back to their good old garage days!" With that display of dwindling faith, he had already started looking out for another major account. And that's exactly how the caviar-laden engineer had shown up at our offices.

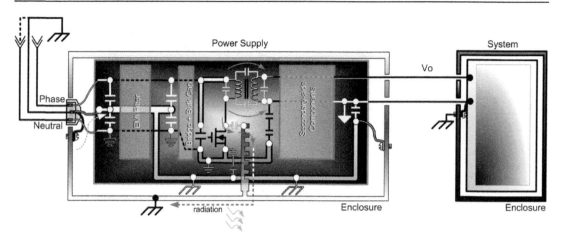

Figure 11-3 Typical Connections of an AC-DC Power Supply

Now you may have been thinking the large metal plate in Figure 11-2 should have somehow *helped*. You are thinking perhaps of a shield. But this is not one! Its effect is almost the opposite. Starting with Figure 11-1, there is a DM (differential mode) noise generator delivering a current of I_{DM}. Assuming the output capacitor is almost ineffective at filtering these noise frequencies, we conclude that almost all the DM current goes out through the output supply cable and then back through the output return cable. There is also a CM (common mode) noise generator, but that cannot deliver any noise current directly into the output, for the simple reason that CM currents need the chassis ground to flow through, and in this case there is virtually none (there are secondary paths for it to flow though). But in Figure 11-2, the situation changes, and so CM current can flow directly. You can argue that an output ripple measurement is only concerned with *differential* signals, for that is what we always pick up with a scope probe. So why would common mode noise have any effect on a (differential) noise and ripple measurement? Unfortunately, in this case, the load is unbalanced, having *unequal impedances* to ground from its two ends (ignore the ceramic capacitor). We have one end of the load directly connected to chassis ground, whereas the other end connects to ground via the load resistor. So the CM current I_{CM} coming from the noise generator divides up *unequally* between the output and return rails, in the inverse ratio of their impedances. But by definition, a CM noise generator is considered common mode only because it produces the *same* currents through *both* the supply and return paths. So in effect, this asymmetric current flow is *a conversion of CM noise to DM noise caused by unequal impedances*. And the scope *will* pick this up!

Here's the math involved, using two examples:

Example: Suppose we measure 2µA going from right to left in one wire. And we measure 5µA going from left to right in the other wire. We don't know the current through the chassis ground (earth). What are the CM and DM components involved?

We have

$$I_1 = \frac{I_{CM}}{2} + I_{DM} = 2\mu A$$

$$I_2 = \frac{I_{CM}}{2} - I_{DM} = -5\mu A$$

Solving these simultaneous equations

$$I_{CM} = -3\mu A$$

$$I_{DM} = 3.5\mu A$$

Which would mean that we have a current of 3µA flowing through earth (chassis ground) connection. And we have 3.5µA flowing in and out of the two wires.

Example: Suppose we measure a current of 2µA going from right to left in one wire, and no current in the other? What are the CM and DM components involved?

We have

$$I_1 = \frac{I_{CM}}{2} + I_{DM} = 2\mu A$$

$$I_2 = \frac{I_{CM}}{2} - I_{DM} = 0\mu A$$

Solving,

$$I_{CM} = 2\mu A$$

$$I_{DM} = 1\mu A$$

We can see that this can be considered part CM, and part DM.

Twisting and Turning Your Way Out of EMI Problems

The way I finally managed to get past the hurdle between Figure 11-1 and Figure 11-2 was by twisting the output cables tightly (see upper half of Figure 11-4). That was a trick we

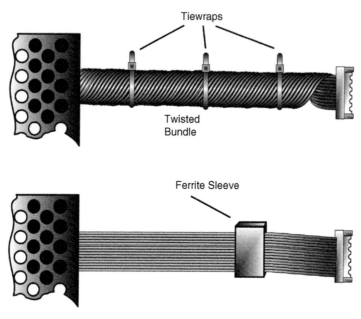

Figure 11-4 Minimizing Noise Along the Output Cable Assembly

had actually learned on an earlier project. Twisting long runs of signal carrying wires is a known way to minimize differential mode noise being picked up through the air, because it increases the capacitive coupling between the cables and also makes both the forward and return cables pick up noise equally (so less noise *differentially*). In our case, though, in principle, twisting can't seem to do much for any common mode noise coming into the cables via *conduction* (from inside the power supply); by helping *re-equalize* the noise picked up (via radiation) along the cable length, it apparently decreases the *converted* "differential mode" component of that noise (caused by unequal impedances). The results actually spoke loudly for themselves. EMI just happens to be a black art at times! Don't believe every so-called expert, because there are very few real ones out there in this particular area of expertise. This is actually one area of Power Conversion where you need *not* be ashamed to admit that you resorted to *symptomatic troubleshooting*! I would be lying if I said I knew all the equations, *and* had a Mathcad spreadsheet to aid me. Baloney, I just try anything to make it work. With a hunch that is proportional to the number of times I received a black eye. I call it a complete hunch of nonsense.

We had gotten roughly the same empirical level of improvement when we tried a ferrite sleeve as shown in the lower arrangement of Figure 11-4. But the sleeve works mainly by increasing the impedance on both lines to the common mode noise coming out of the power supply via conduction. Ferrite sleeves made specifically for EMI suppression purposes also present a high *AC resistance* to noise frequencies, so they not only help prevent the noise from traveling down by presenting a high impedance, they also help dissipate the associated energy and thereby "kill" it. But it is an expensive solution as compared to the *twist and tie wrap* technique we used successfully in very large volumes to our first OEM customer. Though you have to remember the twist and tie wrap was done out of our Bombay factory, where labor rates are low (and humidity high—it was really a sweat shop after all). Also, keep in mind that if you decide that the ferrite sleeve is more suitable for your production environment, you should position it as close to the load end as possible, so it can deal with any noise pickup along the length of the cable, too.

Low-Noise Transformer Construction Techniques

Figure 11-5 reveals two *low-noise construction techniques*, as applied to a typical flyback transformer. We should compare the right-hand electrical schematic with its equivalent winding version to the left. In the discussion below, we note that though transformers with split windings are not being explicitly discussed here, the same principles can be easily extended and applied to them, too. Here are some observations:

- Since the Drain of the Fet is swinging, it is a good idea to keep the corresponding end of the Primary winding buried as deep as possible; that is, it should be the *first layer* to be wound on the bobbin. The outer layers tend to shield the fields emanating from the layers below. For sure, the Drain end of this winding should *not* be adjacent to the safety barrier (the three layers of polyester tape), because the injected noise current is proportional to the net dV/dt across the two "plates" of the parasitic capacitor (formed by the windings on either side of the interface). Since we really cannot reduce the parasitic capacitance much (without adversely impacting the leakage inductance), we should at least try to reduce the relative dV/dt that appears across this interface capacitor by positioning "quiet layers" on either side of it if possible.

- Comparing the diagram on the upper left with its schematic on the upper right, we see that the "start" and "finish" ends of the windings have also been indicated. In particular, all the start ends have been shown with *dots* in the schematic. Note that in a typical production sequence, the coil-winding machine always spins the bobbin in the same direction for every layer and winding placed successively. That makes *all the start ends of the windings magnetically equivalent*, and we can therefore consider *all* of these start ends either as the dotted ends of the electrical schematic, or *all* of them as its non-dotted ends (it doesn't matter). If one dotted end goes high, the other dots also go high at the same moment (as compared to their opposite ends). We then see that from the point of view of the actual *physical proximities involved, every dotted end of a winding automatically falls close to the non-dotted*

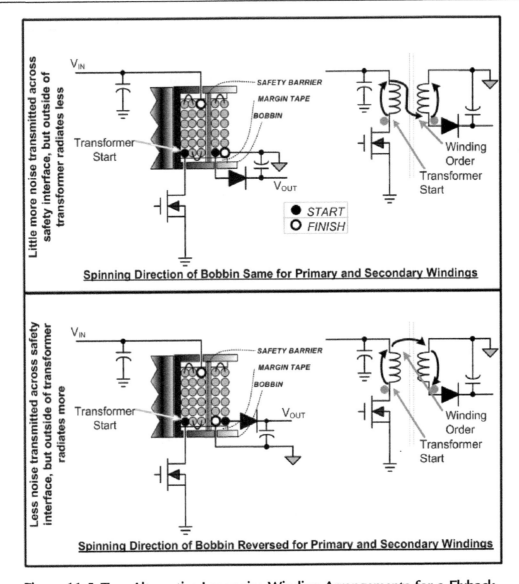

Figure 11-5 Two Alternative Low-noise Winding Arrangements for a Flyback Transformer

end of the next winding. This means that for the flyback transformer in the upper half of Figure 11-5, the diode end of the Secondary winding will *necessarily fall adjacent to the safety barrier* (also called "safety interface" or "isolation boundary"). Yes, because of that we will have a certain amount of dV/dt still present *across* the barrier. But note that this dV/dt is *much smaller than if the Drain end of the Primary winding were brought adjacent to the safety barrier.*

However, the transformer now has the advantage that the "quiet end" (ground) of the Secondary winding is now the *outermost* layer. That is by itself a good shield to prevent radiation from emanating from the transformer. Consider the alternative. Suppose we had wound the transformer the "wrong way," that is, by reversing *all* the start and finish ends shown in the upper half of Figure 11-5. That would have brought the Drain end of the Primary winding right next to the safety barrier, with the Secondary ground end (which is usually connected to the chassis) directly across the isolation boundary. With this winding arrangement, we would have a healthy dose of CM noise injected directly into the chassis/earth.

■ The approach shown in the lower half of Figure 11-5 calls for the winding directions to be reversed in production between Primary and Secondary windings (some call that "unproductionizable!"). But that allows both the quiet ends of the windings to face each other at the safety interface. The amount of CM noise transmitted across the boundary is very low. This is one of the tricks I had evolved for the wish list. The outside of the transformer is then not very quiet, and so a "flux band" is required as shown in Figure 11-6. In the wish list Flyback, I had actually used the same principle but with a foil winding for the 5V/7A output. The 12V/2.5A winding was not of foil, but I had wound it sort of "bifilar" with the 5V foil, so that the foil would shield noise emanating from that winding, too.

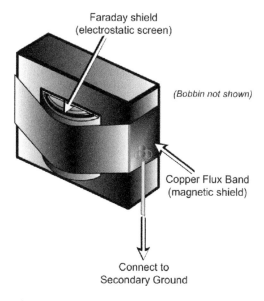

Figure 11-6 A Flux Band to Suppress Radiated Noise from a Transformer

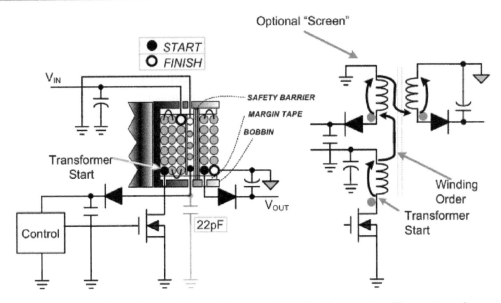

Figure 11-7 Using the Auxiliary Primary-side Winding to Act like a Faraday Shield

Obviously, a few 12V turns were still leftover after the 5V foil winding got completed, and these were simply wound tightly around the inner 5V foil.

■ Another trick I evolved at that time was to use the primary-side IC supply winding to act as a Faraday shield. This is shown in Figure 11-7. Basically, both ends are AC coupled to Primary ground and help sink the noise being capacitively transmitted out of the main Primary winding. Therefore, much less CM noise gets transmitted to the secondary. The technique as shown is applied to the upper half of Figure 11-5, but it can easily be applied to the lower half, too. The disadvantage of using a conventional Faraday shield is that it greatly increases the leakage inductance, and thereby affects the efficiency. That is the reason why we *never* used any (conventional) shields inside our Flyback transformers. Though flux bands we did, *always*.

Location of Clamp Affects Noise, Too

In the do-or-die struggle for that last millivolt reduction in output noise, one of the things I discovered by complete trial and error was that the location of the RCD clamp makes a great deal of difference. In Figure 11-8, we see the two possibilities. Most people prefer the one on the left, because it can be shown that it is less dissipative. But the method on the right side produces the least noise. However, for years I didn't quite figure out the reason for this observation. My best guess now is that it has something to do with the reverse recovery

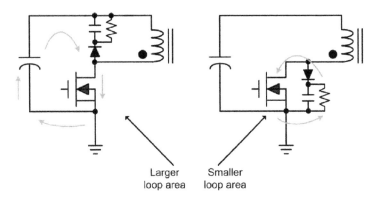

Larger   Smaller
loop area   loop area

Figure 11-8 Location of RCD Clamp Affects Noise, Too

current that flows back through the diode for a very brief moment just after the diode has stopped conducting. The difference is not necessarily in the *amount* of recovery current, but in the *path* it takes. As you can see from Figure 11-8, the path is much shorter if the RCD clamp is ground referenced. In the other case it has to traverse a much larger loop, and that increases the noise emitted.

A Cheat-sheet for EMI Troubleshooting

Here we are going to focus at different parts of the power supply shown in Figure 11-3 and detail a series of things we need to keep in mind. We are assuming this is a single-sided board.

During layout, try to create a fairly thick copper island from the primary side, running very close to the secondary side. This island serves as the PCB-level chassis ground (hereafter called the "PE ground," where PE stands for protective earth). The reason for doing this is that CM noise is created *inside* the power supply, and its generator is essentially a *current source*. So just throwing in an impedance wall to somehow contain the noise will not help, unless we provide an *alternative path* for it to flow through. If we don't, then just like normal inductor current, this noise current too will try to force whatever voltage is necessary to surpass the impedance wall and keep on flowing. Since our ultimate purpose is only to prevent the CM noise from flowing into the *output or the input*, we need to do the following:

a)   We have to *allow* the CM noise currents to circulate *within the power supply*.

b)   We also place high-impedance walls (e.g., common mode chokes) and/or low-impedance shunts (capacitances) at the input and output (if necessary), to contain the CM noise within the power supply.

c) We should try to reduce the *area enclosed* by the circulating CM current loops to prevent radiation.

d) If possible, we also want to deliberately introduce *dissipative elements* (not just parasitics with inherent AC resistance) in the circulating path to kill the energy associated with the CM noise (e.g., lossy beads).

The first allowed level of allowed circulatory paths is on the PCB itself. The second allowed level is through the chassis (enclosure). Though the latter path is inevitable sooner or later, it is not preferred since the currents through it cause the enclosure to radiate H-fields (also E-fields if not earthed/grounded well). Therefore by creating this PE ground island, we hope to connect it by means of "Y-capacitors" to carefully considered points within the supply, both on the primary side and secondary side. So a fairly long and thick copper island serving as the PE ground will help create relatively small circulating noise current loops on the PCB.

What are the positions of Y-capacitors we can think of on the PCB? Of course, an input EMI filter is always present, with DM and CM filter sections as shown in Figure 11-9. Though sometimes, the leakage inductance of the CM choke serves as a sufficient (inadvertent) DM choke, so separate DM chokes may not be present. We can find Y-capacitors in the CM section of any such input EMI filter, connected symmetrically to PE ground (also labeled "E" for earth). Many designers, however, just put this input EMI filter in and think their job is done, and that it will *somehow* work. But they forget they need to *clear the entire way* for the CM noise current to flow. The Y-capacitors in the input EMI filter stage only constitute a small part of that closed path. Also, unlike analog/mixed-signal applications, in a switching power supply the CM noise currents are not generated by some static leakage paths to PE ground, but by the switching action injecting noise through the parasitic capacitances (using the equation $I = C dV/dt$). Behind the switching action is an inductor, and that inductor also forms the driving force behind the CM noise currents.

The main injection of CM noise to the output occurs through the parasitic capacitances in the transformer. Another point of injection is into the chassis via the heatsink of the switch, unless of course the heatsink is *not* connected to the enclosure (in which case it needs to be connected to the Primary-side ground). Since the Secondary ground is ultimately connected to the earthed enclosure, the injected output CM noise will find its way to the input side, too. Note that if we try to limit the CM noise by making the enclosure floating, though the CM noise *currents* through it will certainly be forcibly curtailed (and therefore their associated H-fields), now the *voltage* on the enclosure will start swinging, and it will become a giant E-field radiating antenna. So connecting the enclosure to the earth terminal of the AC inlet is not just necessary for safety reasons, but from an EMI point of view, too. Unfortunately that also provides easy access for some of the CM noise to flow straight through into the wiring of the building. Therefore some people use a "ground choke." This

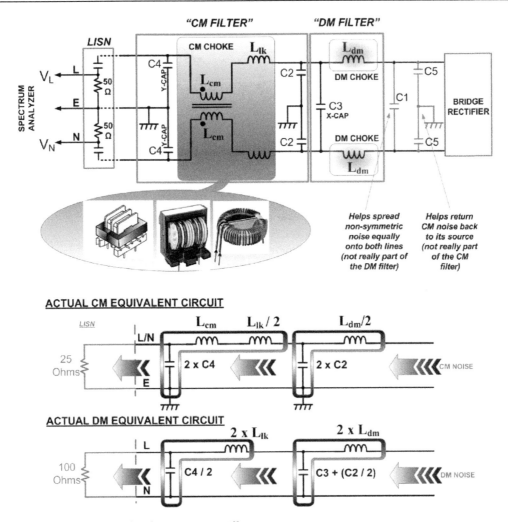

Figure 11-9 Standard Input EMI Filter

is a small ferrite or powdered iron toroid inserted on the wire connecting the earth terminal of the AC inlet and the PE ground trace of the PCB. However, this little choke has been known to cause severe system issues and imbalances, even leading to failures under certain input surge conditions, and should therefore be avoided like the plague. Besides that, though it does often seem to help restrict the CM *conducted* noise appearing at the input, it usually does that at the expense of the *radiated* emissions. You can't win! Not with a ground choke.

Now we will focus on specific areas of the power supply shown in Figure 11-3 and see what the options are.

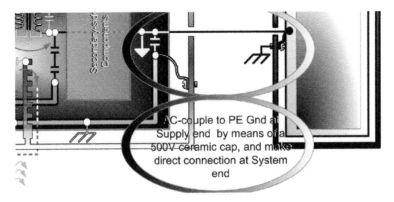

Figure 11-10 AC Coupling of Secondary Ground to Protective Earth at Power Supply End by Means of any 0.1µF Disc Ceramic Capacitor

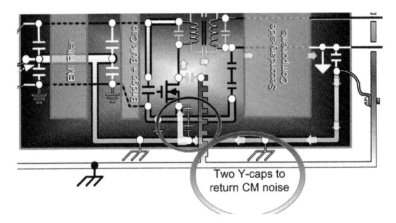

Figure 11-11 Two Safety-approved Y-capacitors from Primary Ground to Protective Earth to Help Complete the CM Noise Loop Close to the Power Stage

Recommendation 1 (see Figure 11-10): To avoid allowing the CM noise to flow down the cables before it closes its loop, place a disc ceramic capacitor between Secondary ground and PE ground. This need not be a safety-approved Y-capacitor.

Recommendation 2 (see Figure 11-11): Two safety-approved Y-capacitors between the PE ground island and the Primary-side ground help return the noise routed into the PE ground trace by other Y-capacitors.

Recommendation 3 (see Figure 11-12): A common EMI trick on commercial power supplies is to place two safety-approved Y-capacitors between the Primary and Secondary grounds. Some suggest a better place for this is from the HVDC rail to Secondary ground. But sometimes

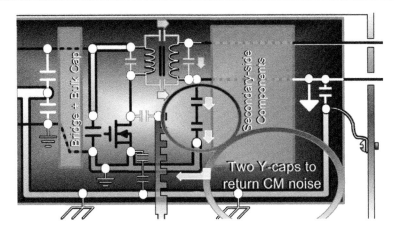

Figure 11-12 Two Safety-approved Y-capacitors from Secondary Ground to Primary Ground to Help Complete the CM Noise Loop Close to the Power Stage

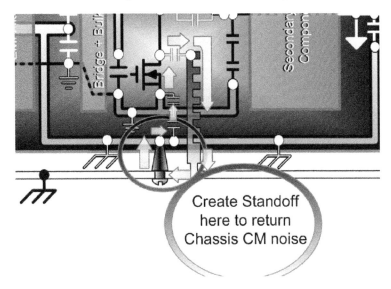

Figure 11-13 Create a Metal Standoff Near the Switch to Close the Loop of the Noise Current Injected Through the Parasitic Mounting Capacitance of the Heatsink

these Y-capacitors only make things worse. It probably depends on which ends of the Primary and Secondary windings are adjacent to each other across the isolation boundary and thereby linked capacitively. There is some trial and error in this, so simply leave provision for alternate mounting approaches (and be prepared to omit these capacitors entirely).

Recommendation 4 (see Figure 11-13): Create a metal standoff near the switch to close the loop of the noise current injected through the parasitic mounting capacitance of the heatsink.

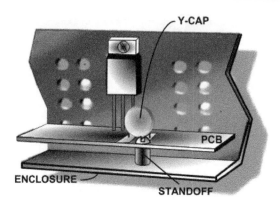

Figure 11-14 The Suggested Metal Standoff Technique to Return the Noise Current Injected Through the Parasitic Capacitance of the Heatsink

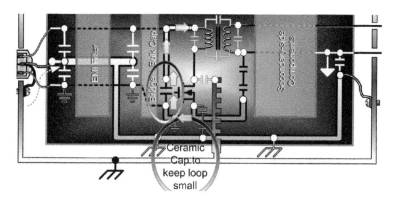

Figure 11-15 A High-voltage Decoupling Ceramic Capacitor Close to the Power Stage Helps Complete the Noise Current Loop

Recommendation 5 (see Figure 11-14): This is the implementation of the above recommendation. Two regular Y-capacitors or one higher-voltage Y1-capacitor will allow the injected CM noise to return to the power stage (where it is being created to start with). Chassis-mounting of the switch for good thermal management is really not as scary as most engineers feel, if this technique is implemented to go along with it.

Recommendation 6 (see Figure 11-15): A lot of the CM noise needs to be pulled through the bulk decoupling capacitor, which may be far away, besides not being a good high-frequency component anyway. So to close the CM current loops as close to the switching stage as possible, a high-frequency ceramic decoupling capacitor between HVDC and Primary ground can help a great deal.

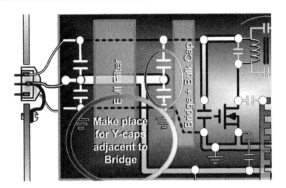

Figure 11-16 Provision for the Possibility of Placing Two Y-capacitors Adjacent to the Bridge Rectifier

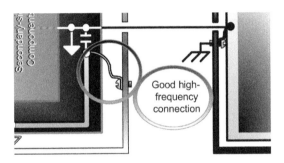

Figure 11-17 Establish a Good High-frequency AC Connection to the Enclosure at the Output-end of Power Supply by Means of Thick Braided Wire

Recommendation 7 (see Figure 11-16): Rather than wait to return the CM currents close to the inlet, it is better to place two Y-capacitors just preceding the bridge rectifier. For this reason, many engineers reverse the order of the CM and DM sections of the input filter. In Figure 11-9, the CM stage is shown closer to the inlet than the DM stage. But that makes the Y-capacitors right next to the inlet. And you would prefer to shunt the CM currents before they get that close to the wiring of the building. So by making the DM stage closer to the inlet, the Y-capacitors of the CM stage move away from the inlet. The only caution to be exercised in placing the DM stage close to the inlet is that this must really be a symmetrical stage. Because any asymmetry will amount to creating a CM component. And unfortunately, there is no CM filter waiting after the DM stage to absorb this any more.

Recommendation 8 (see Figure 11-17): A good high-frequency AC connection between Secondary ground and the enclosure is recommended at the output end. A thick bunch of braided copper helps.

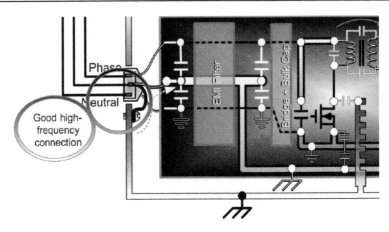

**Figure 11-18 Establish a Good High-frequency AC
Connection to the Enclosure at the Input-end of the Power
Supply by Means of Thick Braided Wire**

Figure 11-19 Never Use a Ground Choke

Recommendation 9 (see Figure 11-18): Similarly, a good high-frequency connection at the input end also helps.

Recommendation 10 (see Figure 11-19): The location where a ground choke is typically inserted (and shouldn't be!) is shown in Figure 11-19.

The final message is, "Be resourceful." For example, we had a lot of luck inserting lossy ferrite beads at critical positions (but check your efficiency afterwards!). Once a particularly severe spike in the EMI spectrum was traced by us to some strange resonance between the Y-capacitor and the traces leading up to it! By raising one end of the Y-capacitor and inserting a small ferrite bead over its lead, I could almost kill that spike. But if I raised both legs and inserted two beads (one on each leg), the *overall* EMI spectrum deteriorated significantly. In any case, this beaded solution didn't "look" good, and to avoid alarming the customer with our unbridled creativity, we finally reworked the PCB, and fixed it for good.

Discussion Forums, Datasheets, and Other Real-World Issues

Thinking Is the Key

While Googling the other day, I came across an interesting online article titled "Power Supply and the *Thinking* Engineer," co-authored in January 2005 by Bob Pease himself. After reading it, I hoped this was only done as a favor to somebody. Because I found it hard to believe that the man well-known for throwing computers off the roofs of Building D in Santa Clara had finally settled down to a day job exhorting engineers to use his company's online software simulation tools (on *computers* around the world of course), specifically for designing and troubleshooting *switching power supplies* (which I know he has no clue about)! The article says, "The company's online tools can be used to discover design problems and correct them—as long as thinking is applied as well." But what about "errors" in the online tools themselves? Who's watching them?

Anyway, *thinking* is what I have always been recommending all along, too, so we are on the same page. But I realized I needed to really *think* this through for *myself*. Because thinking (i.e., analysis) must *follow* a systematic phase of data collection, not precede it. In Chapter 1, I quoted extensively from Ronald Hughes' article. In that he wrote: "Start with fact, end with fact, and what you have is fact, not supposition. . . . Analyze using short deductive steps in logic, and then verifying at every step during the logic development process . . . Data is definitely the key to successful analysis."

If this process is done assiduously, sometimes we might arrive at the *opposite* conclusion we initially foresaw. We may even suddenly realize that we are a part of the very problem we are trying to fix—it could be in our own backyard. Now we really have our work cut out for ourselves! But that's what you call engineering.

Cross-check Everything

Bob Pease himself rather candidly put it, "if you stand on a big soapbox and rant and holler, people will often think you know what you are talking about. They stop looking for mistakes . . . and that's a mistake." But those days are gone for good, thankfully. We all are now subject to intense technical scrutiny from peers. Marketing can't engineer truth anymore.

A few years ago I had just released an application note titled "Stresses in Wide-Input DC-DC Converters." In that was featured an extensive reference design table, the same as the one appearing in the Appendix of this book. Previously, this very App Note had managed to morph itself into a cover story article in the magazine *Power Electronics* (formerly PCIM). I was still terribly nervous—*were all my equations right*? I had tried triple-checking each of them, plugging them into Mathcad for sanity point checks, comparing that against hand calculations, and so on, but you never know!

If you have been around for some time, you would know that we all make horrible mistakes. But some of our mistakes even find their way undetected into major magazines (and books). And then, not only do they sometimes stay that way, but maybe win an award or two, as well! One hilarious example of this is in Figure 12-1. I just happen to know the history

Switched-capacitor IC and reference form elegant -48V to +10V converter

April 24, 2003

A system designer must almost always face a trade-off in choosing the right part for an application. The trade-off usually involves performance, price, and function. An example is the issue of powering circuits from a telecom-voltage source. Telecom systems almost exclusively use high-potential negative rails, such as –48V. Digital circuits typically in use in such applications usually operate from a "brick"-type power supply. However, analog circuits rarely require enough power to justify using a costly brick. At the heart of these bricks is nothing more than a specialized switching converter in tandem with an isolated flyback-transformer coil. But some applications neither require nor can tolerate the use of a coil-based approach. Figure 1 depicts a way to address the problem. The circuit provides a small amount of power to analog/digital circuits, such as the LMH6672 DSL op amp.

The LMV431 voltage reference, along with the voltage-setting resistors sets the output voltage to approximately (1 +1 kΩ / 280 Ω) × 1.24 V ~ – 5.7 V. This output voltage then goes to the base of Q1, the 2N 2222 transistor. The configuration of the transistor causes a VBE drop of approximately 0.7 V, resulting in a net voltage of –5V for the next stage. The purpose of the transistor is to provide additional current to the LM2682 switched-capacitor converter. Note that the converter has a –5V reference (GND pin). Small capacitors C1 and C2 enable the pumping and inverting action required to convert the –5V to 10V. Furthermore, the MSO-8 package of the LM2682 and the SOT-23-3 package of the LMV431 allow the circuit to consume little board space. In roughly the size of a small transformer, the proposed circuit does an elegant job of powering low-power circuits from a negative high-voltage source.

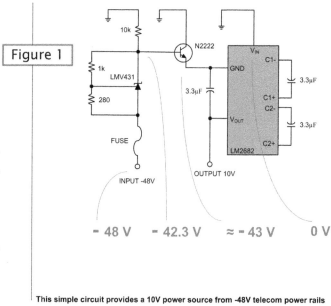

This simple circuit provides a 10V power source from -48V telecom power rails

Figure 12-1 A Glaring Mistake in a Design Idea Published in an Authoritative Magazine for Professionals (W. Ly, Switched-Capacitor IC and Reference Form Elegant −48 to +10V Converter, *EDN*, April 24, 2003)

behind this one, and it makes for interesting reading. One evening my clever colleague showed up in my cube, excited to show me his very latest idea. He was always bristling with great ideas, and this one seemed to be no exception, at least initially. It involved a switched capacitor IC, but I was hopeless at that anyway. However, not to let him down, I looked quickly at the datasheet and realized that the part was simply an *inverter-cum-doubler*. In other words, it would take the $+x$ Volts applied between its Vin and Gnd pins, and produce $-2x$ Volts between the Vout and Gnd pins. Then I started to follow the voltages through as indicated in the blurbs in the figure. The 431 shunt regulator had been set up for 5.7V, so the rest was easy. That's when I turned to my colleague and asked him: "Have you even built this? Does it really work?" He was smart enough to immediately realize he had screwed up somewhere. For one, he was applying 43V between V_{IN} and GND! He gaped at it, and then we both laughed merrily as he admitted he had an entirely different IC in mind for the job. And in fact an entirely different circuit in mind too! Unfortunately, he had jumped the gun already, and this idea had already been sent into the approval process of the company, and also to the editors of the magazine. We decided the best face-saving strategy here was to simply say there was a typo and resubmit it to all concerned. And I know he did that promptly. But what appeared finally in print was the "wrong" (previous) idea. And what's more, at the end of the year, it even found its way into the hallowed 10 or 20 best design ideas published in the magazine during the entire year. My colleague was chuckling away at how that had happened! We just silently hoped and prayed nobody tried to build it! Come to think of it, all submissions are supposedly submitted to industry stalwarts to pick for publication. So maybe some analog expert out there had had one too many that particular evening. Or maybe he was just too busy slamming the daylights out of poor Mr. Taguchi! More likely, he was just an analog guru in the very same company, not a power expert.

The lesson I learned from all this was that we shouldn't outright believe anything put in front of us, even if it is on semi-glossy paper or in high-definition video or Flash HTML format. As engineers, we need to *put pen to paper*, and at least do a sanity check. And lest it be misunderstood, don't forget to do the same for this book too.

Product Liability Concerns

A few years ago, a major U.S. customer of ours suffered a "line down" in their huge projection TV factory in Mexico, all because of a problem with our switcher IC. I was dispatched almost overnight, along with our respected, senior-most QA expert, straight out of our California location (without even proper re-entry paperwork). That was a blistering response time for sure! And we did solve the problem in a couple of days, too, though one particular incident within that period is still fresh in my mind. I was sitting there on the side of the customer's now-deserted production floor, trying feverishly to understand the problem, that is, basically just doing some crisis-mode troubleshooting. I guess I may have been thinking aloud like any immersed engineer, because I was told that our customer's production head had just passed by within earshot. And perhaps he had actually heard something. All I know is that our seasoned QA expert suddenly knocked me out of my engineering semi-trance, with a rather scary, job-threatening stare, "you just admitted to *them* that we have a problem with *our* part. Tell me: are you working for *them* or for *us*?" Whoa! Pardon my naiveté/ignorance, but I thought that was (almost) the same thing.

Basically, our QA expert thought all business was primarily *about product liability*. "Don't admit to anything," was the underlying creed. Later, he told me he had learned this "important lesson" over several years while being mentored by the widely respected Product Engineering manager of our product line. So, all we had to do now was prove that our part worked on our own eval board, with a benign resistor stuck at the end of it, and we were free of *liability*. Of course we would then *help* the customer, *provided we had nothing to lose*. As it turned out, our part *was* actually producing a severe overshoot at startup, thus causing damage to their system board. But as a last eventuality, we were even prepared to argue endlessly that it was within spec, for the simple reason that nowhere in our datasheet did we ever specify how much overshoot our part could have. So if you think that an overshoot of 1V for a 5V output was unacceptable, that was just *your* problem! I have talked about such implied expectations in Chapter 8. Get these clarifications in writing from the vendor before you buy. This can really be their response when all their chips are down.

You should always be very careful what you *think* the datasheet says, as opposed to what it *really* does.

Concluding the Mexican adventure, by examining the stamped date codes, we finally traced the problem to some manufacturing tolerances on our switcher IC. We also then created an Apps fix (in this case, a soft-start tweak if I remember right), and we were on our way back, *with no liability*. That was of utmost importance.

So if you think your vendor is stonewalling you (when you ask some potentially embarrassing questions), that could be because *he or she is*!

It's All about the Customer

To write this particular chapter, I was getting prepared to painfully sift through, collect, and analyze any relevant data available. I had told myself I was not going to be the one sitting in an ivory tower blindly *guessing* what the *real-world* issues are. This really had to be as real as it gets. So in my ensuing search for information (data collection phase), I chanced upon a popular online discussion forum where real engineers talk about the actual issues plaguing them concerning switching power supplies. Aha! What better repository of real-world problems can there be than that? Though I realized that that particular site was started by an analog company with the intention of supporting its specific products, I also realized that if you take the trouble of looking past their part numbers, you would arrive at the crux of the matter—the *engineering* aspects of some of the key issues being faced in switching power supplies in general.

The very idea of a company starting a forum such as this one is essentially brilliant and thoroughly laudable. It also imparts a *perception* of transparency to their operations from the get-go. But as mentioned, being a die-hard engineer at heart, I first needed to collect and analyze the emerging data, before reaching any conclusions. I delved deep into their forum, and also followed up on their directions and suggestions, to see where that would lead me. But slowly a sense of shock started creeping in. I saw that many questions had been answered very inadequately, sometimes peremptorily, and sometimes even outrageously. I therefore decided to add my own detailed analysis, for whatever it is worth, on the pages of this book. There was so much to add to their responses, I could only take up less than twenty questions here, though I was planning more like fifty.

Worse, having been in the business for many years (they too!), I could sense at what point they were just trying to slip and slide their way out of embarrassing questions (liability concerns?). To me, that's akin to bolting from the scene of an unintentional car crash, that is, an accident made into a hit-and-run situation. The problem is *not* that there is "bad part" out there. *Nobody* has a perfect product (though certainly, there are degrees of acceptable imperfection too that I will not debate here). I feel that a company trying to *appear* to have the highest levels of transparency also needs to be able to accept serious mistakes or criticism without hesitation (transparently!), and then, if necessary, ignoring the short-term motivation of the all-important balance sheet, immediately recall a product that they now realize a whole generation of engineers is struggling hopelessly with. I know of companies that did that, winning great *long-term* respect and loyalty from their customers. At the end of the day, it is all about doing business *with customers*. We must work in *their* interests ultimately, or there is no business left. Never mind all the QA experts of this world and their sleazy mentors. There must be some difference left between us and used car salesmen.

I have been saying this for several years now. And I am actually getting quite used to the fact that usually such advice either falls on deaf ears or drips off some perpetually deadpan faces. But sometimes I do get lucky. After all, change does require persistence. However, back in early 2004, I had been really struggling for months to get my company to guide the customers correctly about a certain problem that was now plainly getting bigger and bigger by the minute. I had just sent out a rather firm but plaintive Email to all the first-level managers of our product line about how many actual customers had started complaining specifically about that issue (three in the past two days alone, to me personally), and that we really needed to come clean now, or at least to fix the problem posthaste at the design level once and for all. That Email was again virtually blank-faced (and to this day the problem still lies unattended to). But refreshingly, a young and brilliant engineer-cum-software-expert, working on the same project (let's call him "Ben"), apparently without the slightest hesitation, wrote fearlessly back to me on the company Email:

(**March 25, 2004**) I agree with you completely that the problem should be addressed at the design level. This type of thing is the reason why some of our customers perceive us as not addressing problems at the root cause. I think it is ridiculous that we are expecting customers [to somehow] accommodate a device that we are [actually] selling TO THEM.

Unfortunately, I heard that in early 2006, Ben finally quit the company, electronics, and even software for good. What a loss! It really doesn't pay to preach the gospel to everyone.

The Q&A Session

I admit the following list of questions and answers is not even close to being comprehensive, but I hope it at least reveals how an engineer's *thought process* should proceed under such circumstances. All effort has been made to clean up most of the typos/ grammar of the original forum, and also to improve the language somewhat, for better readability, although for the questions and blog entries, the original typography (no italics, no subscripts, and u is used for μ) has been used, to match the discussion board context. Any possible revelation of identities has been deliberately suppressed, even though this is, after all, a completely public online forum that anyone can access (and add to). Yes, I did once add my two cents worth on the forum directly (but with little success). Also, the part numbers I mention in this chapter should not be considered in any way real—they are there only for the purposes of this technical analysis and discussion. Further, any errors in the analysis or logic are entirely mine. In any case, you should always thoroughly double-check what I or anyone else is saying, before inferring anything whatsoever.

Question 1

April 2000   I am using a pair of 2675's on a board. One to provide 5V for Analog/Digital. The other to provide 5V for a lamp, which blinks at 1Hz and draws 250mA. When the lamp is off the 2675 is in discontinuous mode. When the lamp turns on, the power rail sometimes dips 2V for 100ms before "catching back up." Any ideas what is going on? Would I be better if I applied a minimum load of 100mA at all times?

> *Author's Comments:* The clues here seem to be a) one load is a *lamp* b) that this happens only "sometimes," and c) that the output rail dips for 100ms (but which one?). We should ask.

Blog Entry 1   When the load is removed, such as when the lamp blinks off, the 2675 goes into discontinuous mode because of the zero amp load current. Since it takes 1 second for the lamp to blink ON (load application), the energy previously stored in the capacitor is discharged through the catch diode. On reapplication of the load, it takes time for the output voltage to ramp up, which explains the transient "dip" of approximately 100ms you observed. The idea of connecting a constant load should cure this problem as you speculated. A resistive load (250 ohms) drawing a minimum of 20mA should suffice to maintain the specified load regulation.

> *Author's Comments:* The above official explanation seems to suggest that the load comes on *slowly* over one full second (though the customer has only said the blinking *rate* is 1Hz). And that it fully discharges the output capacitor "through the catch diode" (which implies the switch is *not* turning ON for some strange reason, despite the fact that the output has dipped severely). In fact only when the load blinks back OFF and then ON again, the converter *suddenly realizes* what is expected of it all along. But, unfortunately, it is too late and it has to virtually start from scratch to rebuild the energy in the discharged capacitor. Well, that is the official explanation (or best guess) anyway. Yes, voltage-mode converters do recover somewhat slowly in going from minimum load (~0A) to maximum load because they have to pass from DCM to CCM, in which the power stage gain changes from a single-pole response to a double-pole response. But 100ms is a very, very long recovery time for any modern high-frequency switcher. It is equivalent to a 10Hz waveform! It really doesn't make sense! Further, the load mentioned by the customer is 250mA (this being a 1A switcher), whereas the official explanation calls out for a minimum load requirement of 250Ω, that is, 20mA. But there is no stated reason to explain why "20mA should suffice." Why not 40mA, or 10mA? All Non-Synchronous converters go into DCM at light loads, but in a typical design where the inductance is selected for a current ripple of ±20% at max load, the CM to DCM transition occurs at 20% of maximum load. So that would take us to one-fifth of 250mA or 50mA, not 20mA. In other words, 20mA would not usually be enough to ensure CCM (unless the inductor is very big, but the customer was *not* queried about that). And therefore, *if* the DCM to CCM transition was the root cause, as the official believes, even his or her suggested solution of 20mA minimum load would not work.

Blog Entry 2   At 250mA, your lamp is unlikely to be solid-state. Incandescent lamps may draw considerably more current than you think at startup, if the element temperature is not at thermal equilibrium. Though one Hertz is fairly fast for thermal effects, it is worth keeping an eye on.

End of Thread.

> *Author's Comments:* Yes, at a rate of 1Hz, we can safely assume that the cold resistance of incandescent lamps is no longer an issue—the filament would already be hot, having no time to cool down sufficiently between successive blinks.
>
> It looks as if no one even considered how the improper paralleling of two converters of the same supply rail can cause issues because of their mutual interaction. This issue was discussed in Chapter 8. To me, that is the main suspect here. I would have asked the customer a simple question—how does the problem respond to completely disconnecting one converter? If the answer to that is that the problem persists, I would immediately suspect the PCB layout.

Question 2

April 2000   I am using the 2678-5.0 in the following configuration—Input voltage: 15 to 34V; Input Cap—3 × Kemet Series T495 4.7uF/50V; Inductor—Pulse PE54041 22mH; Diode—MBR745; Output Cap: 3 × AVX series TPS 100uF/10V; Iload—75mA to 2.8A. Problem 1) At power-up, Vout rises to 5V, when Vin reaches approx. 8V with a rise time of 3V/25uSec, then Vout peaks at 6V (5Vdc + 1Vpeak), why? Problem 2) If I disconnect the Load (I = 2.6A), then Vout rises excessively, and then with a big noise and flames, the output caps are destroyed, why?

Blog Entry 1   Sorry it took so long to get back to you. I wanted to look at the question in Problem 1 below in more detail. First I'll answer the question in Problem 2 since that one is more obvious.

The 267x regulators have a capacitor boost voltage circuit in them (which uses the capacitor Cboot) to give the power switch increased Gate drive and therefore higher efficiency. However, to protect the power switch there is a stack of two 5V zeners so that the Gate voltage will not be more than 10V above the source. Because of this, when the input voltage is 10V or more above the output voltage the zeners can break down and some current can leak through. This current is very small, but if there is no load on the output it can still cause the output capacitor to charge to higher voltages and in this case, blow up the Tantalum capacitors. This is easily fixed by putting a minimum load on the output to draw away this leakage. 1mA (5k resistor) will be plenty to ensure that the output never rises with no external load connected.

As for the output overshooting at startup, this is harder to figure out. I looked at this for some time in the lab with a board I built. I used the same power components as listed below, and I used other values as well. I tried duplicating the input slew rate as well as varying it and varying other conditions. I could not get my board to overshoot. So there are a few thoughts I have on what it might be, but I can't prove any one of them. There is the possibility that this power source is different in some way. The slow slew rate you are using may be affecting the startup. I tried similar slew rates, but the differences in how the supply handles it could affect it somehow. The small input capacitance may be worsening the effects. If I read it correctly, there is only about 15uF total input capacitance there. Although I can't say for sure what negative effect this might be having in this case, or even if there is one, I would recommend increasing this capacitance to at least triple this value. But I think the input ramp should be checked to make sure the input is not hitting a plateau when the current draw begins. If it were to plateau (possibly current limiting the source), the output may be allowed to charge quite high before the control circuitry has enough voltage to begin working. Another possibility is the inductor being used. If its saturation current rating is very low, it may be saturating during startup causing even higher currents. The high current and energy stored could cause some overshoot. When the output overshoots (for whatever reason) the regulator can only respond in a certain amount of time, limited by its bandwidth. So increasing the bandwidth of the regulator by lowering the inductor value and/or output capacitance (pending stability, should be checked on 267x made simple) should help reduce the overshoot and its duration.

I hope I explained this well enough to be understood. Basically Problem 2 is easily defined and solved, whereas Problem 1 is not. As I have mentioned, I have not been able to reproduce this overshoot on the boards I have. The best I can do about this now is to have you check the issues I listed above. Please let me know if you have any other questions or if any of this is confusing or poorly explained. Thanks.

Blog Entry 2   Hello again. I think that my question was poorly formulated, sorry. I get your point, thanks. I have analyzed my problem better, I think!! (The component values are from your company's own design software.) My problem is in relation to a blowing fuse. I just have a 2AT fuse in front, between the main 24VDC power supply and my circuit. And when I apply a heavy load of 3A (the idle current is approximately 1A), the fuse blows and then my problems start. The output is also oscillating with a high input voltage. Can it be because the device has a problem regulating when the input voltage drops fast because of the small amount of energy in the input cap?

End of Thread.

> *Author's Comments:* This is probably a 2A fuse with "T"—that is, Time Delay. The time delay ensures that the fuse will not blow due to the high inrush current of most switchers. However, also note that the max load being applied is 3A. But this is a 24V to 5V conversion, so the input supply current will be almost 24/5 ~ 5 times less, that is, only *0.6A* for a 3A load, not even close to 2A. That cannot be the issue. In any case, there is almost no way the input fuse can blow if the switch of the IC has not failed. A blown fuse in power supplies is almost 99.9999% the *result* of a *switch failure*, caused by some other malfunction that we need to investigate. Further, no problems can *start* once the fuse blows, as the customer seems to state! That seems like an obvious typo, but it should have been clarified delicately.
>
> In general, output overshoots can be the result of several causes, such as:
>
> a)   A very stable, but *slow* loop (too low a crossover frequency and/or too high a phase margin leading to sloppy correction—"I know I am off the mark, but I will take my own sweet time to get back").
>
> b)   A very *fast* loop (too high a crossover frequency and/or too low a phase margin, causing excessively fast correction that tends to overshoot on to the other side, then correct again, etc., leading to output oscillations/ringing and sometimes even complete instability).
>
> c)   Going from 0A to full load can always be a bit of a problem because of the DCM to CCM transition mentioned previously. In current mode control, the power stage has a single-pole response even in CCM, so the problem is not considered as acute as in the case of voltage-mode control. Incidentally, there are separate camps of devout followers swearing either by current mode control or voltage-mode control. At last count, voltage mode seems to be winning once again (when combined with line feedforward techniques).
>
> d)   Note that a 0A-to-full-load step (or back) is not really considered a small-signal loop issue anyway. It is really a *large*-signal response issue. The biggest impact on the undershoot/overshoot observed in this case comes simply from the amount of bulk capacitance (and stored energy) present at the output. Because in the initial instant of sudden application of load, the bulk capacitor (C_{OUT}) alone provides the energy being demanded by the load, until the loop finally kicks in to prop up the falling rail. Note that it takes several cycles for the current to ramp up to the new required level in the inductor. So small inductances tend to help in quick recovery and help achieve a fast loop response (of course provided they don't create full-blown instability in the process).

e) In large signal responses, the error amplifier can easily "rail" (its output hitting either of its supply rails), and this could saturate its internal transistors. These may take time to recover and correct the duty cycle. The 267x family has an innovative *two*-stage error amplifier, but this also unfortunately severely exacerbates the slow recovery once the two error amplifiers have railed (as under severe load steps).

f) Many soft-start circuits take control of the reference voltage applied to the error amplifier in the initial few milliseconds of power-up, thereby causing it to ramp up slowly (along with the duty cycle). But at some stage, they need to *hand over* control back to the fixed internal bandgap reference. In doing so, there is always some offset error in the process, and this makes itself felt as a glitch/spike (undershoot or overshoot) at the output. Though this output perturbation is usually small and often tends to be absorbed by the output bulk capacitance, you should always watch out for it. Particularly because in our case, this hand-over problem could well be an underlying issue with the 267x family itself. Note that even the datasheets of these devices admit that (severe) overshoots can occur at startup. And the amount of overshoot is significant when there is *no* soft-start capacitor present (reflecting an inability of the *loop* to respond fast enough to the suddenly rising output—a fairly common situation in many ICs), but also if you have *too large* a soft-start capacitor. Unfortunately, the 2679 datasheet only carries this soft-start versus overshoot warning on page 12 of its datasheet. And sadly, the 2678 being discussed here has no soft-start pin available anyway. Also remember, that soft-start *usually* only works in the initial power-up sequence. Once you are already powered up, and then if you short the output and release it, there will be *no soft-start present* to help the recovering output rail. So if the IC has any architectural weaknesses, you may get to see these simply by shorting and releasing. Which is why soft-start may be a nice thing to have in the long run, but *if you are evaluating a given part, remove soft-start and then check for output overshoots*. This is the peeling of the onion approach I have mentioned in many places in the book. Note that there are ICs out there that *reinitialize* after a fault. So if the output is shorted and released, a comparator detects the output undervoltage and then proceeds to discharge the soft-start capacitor fully before starting up again. That maintains soft-start even under fault conditions. So in such cases, you do *not* need to test the part for overshoots with the soft-start capacitor removed. I personally like this soft-down scheme, but in certain applications, it can be perceived as a nuisance (though it is certainly safe).

g) It is true that even if you turn the Fet OFF completely, there is always some leakage current that continues to flow through it. And this can get worse as the input voltage is raised (and also at higher temperatures). Similarly, *driver* stages, and in particular *floating* drivers, produce their own intrinsic leakage currents. Therefore, at any given moment, there can always be some unintended current flowing out of the SW pin of Buck ICs. It is usually very small, and therefore, extraneous resistors in the output area (like the voltage divider, or the parallel parasitic resistor across the output capacitor) help sink this current. The capacitor does not keep charging up, and so the user rarely notices. But in many cases, this SW leakage issue can become a problem, demanding the user place a few 100μA to 2mA of preload deliberately at the output. In such cases, if this preload is removed, and if you wait long enough, the output capacitor voltage can keep rising and rising. Note that the feedback loop can do nothing about it (it has already commanded the switch to turn OFF). *Maybe* at some stage this escalating output voltage can cause an output Tantalum capacitor to *suddenly* explode (though I would rather think that in such a case, its dielectric would likely just *barely* break down—enough only to allow the few extra mA of leakage current to flow through, thereby reaching an uneasy equilibrium *without* having to cause the capacitor to *erupt*). But in any case, this behavior is not really considered an "overshoot" because that word typically implies a certain *recoverability*—that is, something *temporary* in nature.

Question 3

September 2000   Hi, I have been designing my power supply with the 2595, in continuous mode of operation for 1A of load current. I am expecting that my load can drop to 5mA at some points during the operation. It may stay at 5mA for a while. What effect will the low load have on my output voltage ripple, and can the output voltage oscillate?

Blog Entry 1   The 2595 is designed to provide +/–4% line and load regulation of the output voltage throughout the range of 0.1A to 1A load current. The regulator will not operate efficiently below 100mA and may become unstable. For the 5V output option and assuming 10V to 20V input variation, the ripple voltage to be expected should range from 25.1mV(min) to 44.4mV(max.). To ensure proper operation of the 2595, even with a 5mA load, an auxiliary resistive load may be connected at the output terminal to draw at least 95mA.

End of Thread.

> *Author's Comments:* The nervous customer is basically asking if all hell will break lose if the 1A Buck switcher is operated at very light loads. The company official is reassuring him or her that there really does need to be at least 100mA always present at the output. That is not true. Every switcher can work up to very light loads. Note that in Question 1, the customer was being advised to load his or her 1A switcher to a minimum of 20mA. In Question 2, it was a minimum of 1mA. Here it is 100mA! It seems to me that this particular advice was simply a question of *two wrongs trying to make a right out of themselves*. Apparently, the support software of the company does not have the *equations for DCM* modeled into it. The obvious conclusion is that therefore all switchers must henceforth be operated in CCM! I am now wondering if this customer ever built this hilarious "100% CCM converter."

Question 4

March 2001   I want to use a step-up regulator (using the 2621) from 3.1V to 3.8V. The problem is the high Vout ripple (100mV). Is there any chance to reduce it?

Blog Entry 1   There are two ways to reduce the output ripple in Boost regulation:

1. The value of the inductor may be increased above the minimum recommended value to reduce input and output ripple. However, once the ripple current is less than ±20% of the average current in the inductor, the benefit to output ripple becomes minimal.

2. The equivalent series resistance (ESR) and equivalent series inductance (ESL) of the output capacitor substantially control the output ripple. Use an output capacitor with low ESR and ESL. Surface mount Tantalums, surface mount polymer electrolytic and polymer electrolytic and polymer Tantalum, Sanyo OS-CON, or multilayer ceramic capacitors are recommended. Electrolytic capacitors are not

recommended for temperatures below −25°C since their ESR rises considerably at cold temperatures. A Tantalum capacitor is preferred since it has a much better ESR spec at cold temperature.

Blog Entry 2   The 2621 uses a comparator with 30–40mV of hysteresis at the feedback pin (1.24V) to act as a trigger to the internal oscillator. If the output voltage is low, the switching action starts. Once the output voltage rises above the hysteretic value, the switching action stops. For an output voltage of 3.8V the ripple caused just by the hysteresis is 122mV. Reducing the ESR or increasing the inductor value will have no effect on this ripple (as it would have in a typical PWM regulator). However, the capacitor listed as Cf is used as a "speed-up" capacitor. The dV/dt noise on the output is injected through this capacitor and the low side feedback resistor to create a higher voltage at the feedback pin to trip the comparator early. You should be able to get the ripple on the output <80mV and as low as 60mV. In fact, if you use a low ESR capacitor on the output, the dV/dt will be eliminated and the output ripple will actually increase. All this I have just explained is not well documented in the datasheet. Note that the reason for this architecture vs. PWM is to reduce the Iq at light loads.

Blog Entry 3   But how can I do it? You don't say.

End of Thread.

> *Author's Comments:* The first company representative seems to have dished out the stock reply on reducing output ripple in conventional (PWM) regulators. But his second point is a fairly good run-down on capacitors in general. However, the 2621 is a *hysteretic* controller and therefore *depends on output ripple* to behave. So the advice doesn't really apply.
>
> We should also examine his first statement more closely: "once the ripple current is less than 20% of the average current in the inductor, the benefit to output ripple becomes minimal." That's not true for two reasons. First, don't forget that the optimum value for inductor ripple is ±20%, not 20%. That's one of the common mistakes I mentioned in Chapter 1 that engineers tend to make almost involuntarily. When you plug in the numbers, you may get the wrong results if you are not very careful. Second, by reducing the current ripple (i.e., increasing the inductance), the benefit to *output voltage ripple* (which is essentially the current ripple multiplied by the ESR of the output cap), is certainly still there (in fact proportional), the only problem being that ±20% forms a point of *diminishing returns*. A current ripple lower than that calls for the *size* of the inductance (its $1/2 \times L \times I_{PK}^2$ rating) to increase almost exponentially; otherwise it would saturate. There is also hardly any further benefit in terms of reduced ripple current through the input and output capacitors. Take a look at Figure 12-2, and remember that though the output RMS current, in particular, decreases proportionally with r in term of percentage, its absolute value is usually very small to lead to any further reduction in the size of the output capacitor. Therefore, an r of 0.4 (i.e., for a current ripple of ±20%) is rightly placed around the "knee," and therefore considered the most optimum.
>
> Looking at the blog entry that follows, this one is technically correct if rather peremptorily explained. Hysteretic controllers are actually quite simple. In Chapter 2 we discussed the *leverage factor* of a voltage divider. In this case the reference is 1.24V and V_{OUT} is 3.8V. So the leverage factor is 3.8/1.24 = 3.0645. So

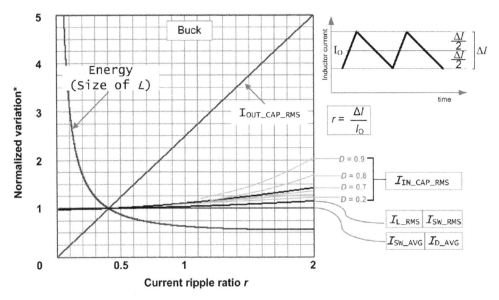

Figure 12-2 How the Current Ripple Ratio Affects Other Parameters in a Buck Converter

if the regulator has 40mV of hysteresis at the feedback pin, this too gets leveraged to $40 \times 3.0645 = 122$mV at V_{OUT}. However, if you bypass the upper resistor of the divider with a "speed-up cap," (capacitor) then from the AC point of view, your leverage factor is now approaching unity. In other words, if you have 40mV on the feedback pin, you will have almost 40mV on V_{OUT}. That is the best you can get in principle, and for that you need a fairly large speed-up cap. In most practical cases, you can get the output ripple close to about 60 to 80mV. But that is it. So the customer needs to have just enough ESR of output capacitor to give about 20 to 30mV of output ripple at the expected switching frequency, and inject it through the speed-up cap on to the feedback pin, along with the DC information coming from the middle of the divider. But what if he or she plans on creating enough ESR to produce a higher ripple, say 120mV? That's not possible, because this is a *self-oscillating* system. So if the customer increases the ESR more than necessary, the frequency will start increasing automatically, so as to continue to satisfy the basic equation $V_{OUT\_RIPPLE} = 40$mV $\times$ leverage factor. In other words, ESR does not matter much anymore, nor the inductance. All that matters are the resistive divider values and the speed-up cap. *The variable involved in this case is the switching frequency.* Of course, the grand prize waiting at the end of the rainbow (with most hysteretic ICs) is a generously laden pot of noise with overflowing jitter. So if you care so much about noise, you shouldn't be thinking about hysteretics in the first place! They should have said that clearly.

Question 5

January 2001 I used the virtual bench design tool to verify my comp selection for 12V to 23V at 500mA step-up converter using the 2577. The virtual bench tool came back with a much smaller inductor (68uH) than the charts and equations in the datasheet seem to indicate. What's the deal?

Blog Entry 1 The inductor size (68uH) determined by virtual bench should take precedence over the value recommended in the datasheet. The virtual bench solution calculates the value of the inductor more precisely, correct to the next lower or higher standard inductor, corresponding to the maximum load current specified. Whereas, in the datasheet, the inductor size is chosen in accordance with a given range of load currents. Generally, the value given in the datasheet is larger. This is to ensure that there is enough inductance for energy storage such that when the switch is OFF, sufficient inductor current flows to charge the output capacitor and supply current to the load.

Blog Entry 2 I'm using the 2622 per Figure 3 of the datasheet. How do I calculate the average and peak inductor currents?

Blog Entry 3 Attached is a file that provides the calculation rationale for the inductor in a Boost regulator design.

End of Thread.

> *Author's Comments:* Yes, I checked too, and the datasheet asked for (hold your breath)—470μH. In fact I ran their virtual bench software, and now it asks for only 56μH—so really, what's the deal? I decided to use Mathcad with the reference equations table provided in the Appendix of this book to double-check. I know that these equations have been checked innumerable times, not only by several engineers, but by several unusually thorough customers, too, over the last several years. Nevertheless, I also decided to compare my results as shown in Figure 12-3 against Dr. Ridley's well-known tool called "Power 4-5-6," and the match was good. The first thing I noticed is that the inductance chosen by the online tool (56μH) leads to a current ripple ratio (r) of almost 2, that is, a ripple of ±100%. The duty cycle is about 50%, so the average inductor current is $I_O/(1 - D) \approx$ 1A. The peak is clearly set unusually high at about 2A. However, the datasheet of the part reports that the inductor current ripple should be typically around "20 to 30%." Note that it does not say "±20 to ±30%." So, in effect, it seems to be asking for a ripple of "±10 to ±15%." That is somewhat unusual, because most people set it at around ±20% (as explained in my book, *Switching Power Supplies A to Z* and also with reference to Figure 12-2). In fact the equations in the datasheet choose inductance based on only ±5% ripple. Alternatively expressed, the datasheet is asking for an r of 0.1 to 0.3, whereas an r of 0.4 to 0.5 is usually considered preferable. But the online tool has just asked for an exorbitant r of 2! And the company official seems to swear by that tool. I notice that to go from an r of 0.2 to an r of 2, the inductance needs to increase *10 times*, that is, 56μH to 560μH. And that is close to what the datasheet is asking for. But read the official explanation again (this time with a chuckle): "The virtual bench solution calculates the value of the inductor more precisely, correct to the next lower or higher standard inductor, corresponding to the maximum load current specified. Whereas, in the datasheet, the inductor size is chosen in accordance with a given range of load currents." If you understood that, please write back to me ASAP.

So why does virtual bench try to place the regulator in almost discontinuous mode? There *could* actually be a good reason for that, since Boost regulators can suffer from a *reverse recovery spike* through the diode when the switch turns ON. Since reverse recovery current depends on how much current the diode was previously conducting in a forward direction, by operating the converter in near-DCM, we can ensure the diode has almost no forward current through it when the switch turns ON. So then there is also no reverse recovery current. This can help in achieving higher efficiency, but you have to check you are not hitting the

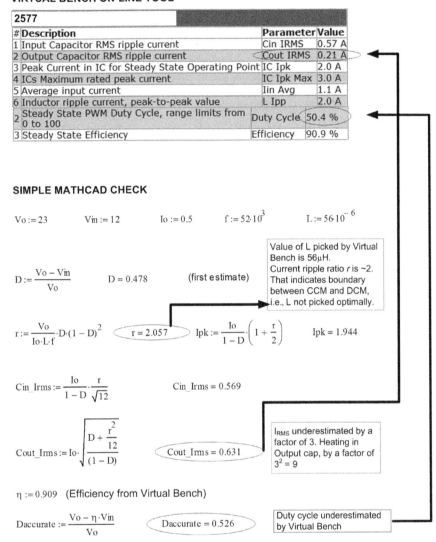

Figure 12-3 A Check of the Results Obtained from an Online Tool for the Boost Topology

current limit in the process; otherwise the output will fold back. The switch current limit is 3A for this part so that is not an issue here. But it calls for *Schottky* external diodes, which in principle have *zero* reverse recovery current. Reverse recovery current just cannot be the reason for suggesting DCM here. OK, another possible reason could be that Boost (or Buck-Boost) topologies in CCM have a *right half plane zero* issue—the "RHP zero"—and if the slope compensation is inadequate they can break up into oscillations. But this part has a corrective ramp, as declared in its block diagram. And besides, the RHP zero is a progressively increasing concern *only as D starts to exceed 0.5.* In this case *D* is very close to 0.5. So ultimately, the logic of running this particular Boost application in DCM escapes me. Unless, of course, there is something else they are not telling me.

I also then checked their virtual bench prediction of the *RMS current in the output capacitor*! This predicts 0.21A as against the double-checked value of 0.63A (see Figure 12-3). This apparent blip in the online tool just happened to catch my attention, mainly because most people know that in a Boost, all the current into the output capacitor comes in pulses through the diode, unlike a Buck regulator where the output capacitor current comes smoothly through the inductor. So the output RMS current of a Boost (and a Buck-Boost) is usually very high and is a determining factor in the selection of the output capacitor. I was therefore quite surprised to see it was stated that low, especially because now, the ripple was also set so high. However, compounding the error, the virtual bench tool also offers you several electrolytic capacitor choices, all of them with an RMS current rating of less than 0.5A, including one with a rating of 0.22A! To be fair, it is at least in line with their erroneous calculations. But I would recommend that you also read the section on the life expectancy of aluminum electrolytic capacitors in Chapter 4. Further, consider the fact that the efficiency prediction of this web tool would also be off the mark, simply because a major chunk of the loss was underestimated.

I started hoping that at least they had got their *Buck* software right. That comparison is shown in Figure 12-4. As you can see, they screwed up parts of this too. *Always take the trouble of validating any calculation or prediction.* Of course corporations have prepared themselves well in advance by slipping disclaimers into every nook and corner. The least you can do is *double-check* and thereby protect yourself from making *non-switching* supplies. It's funny though, they don't spend as much time fixing their problems.

Also look at Figure 12-5 where I show something very fundamental. Whatever you do, you can't break these *three relationships* (one for each topology). For example, if the equation for the Buck were not right, it would imply that the load current and the inductor current are *not* one and the same thing. Which in turn would imply a continuous DC current flowing through the output capacitor (where else would it go?). Which in turn implies you do not have a steady state, and therefore in all probability, no switching converter either! What you really need to do is to start with a first estimate of duty cycle assuming no parasitic drops, then use it to make the first prediction of the currents, then use that to predict duty cycle again, then again predict the currents, and so on until the solution converges. When it does, you will find that the duty cycle and efficiency *will* obey the relationships shown in Figure 12-5. Thereafter, if you have not forgotten any loss terms, your efficiency prediction (and duty cycle) will match what you see on the bench. Sure this takes computation time (and a Mathcad or C++ routine). But you can at least *cheat* your way to the last stage by using the very first estimate of loss and efficiency to at least calculate the final duty cycle using the equations in Figure 12-5, or use your last estimate of duty cycle to recalculate the efficiency. That could help, for example, in finding out whether you may be on the verge of *folding back* because you are too close to the maximum duty cycle limit of the controller.

I was naturally wondering by now whether their online tool at least got these fundamental equations right, that is, whether their efficiency predictions (right or wrong) at least *tie up* with their final prediction of the duty cycles. Unfortunately they do not! They seem to have underestimated the duty cycle (or overestimated the efficiency) consistently across every single part that their tool supports. This can have a major impact on your circuit design and yield. The software will make you think you are still OK, and you will go into production, whereas, in reality, on several power supplies your output could be folding back at that very moment. Also, if the duty cycle is actually more than estimated, your on-time is greater, and the peak currents are higher. So operating close to maximum load, you may even start hitting the current limit of the

TRIAL: 12V to 5V @ 2A

VIRTUAL BENCH ON-LINE TOOL **SIMPLE MATHCAD CHECK**

5005

#	Description	Parameter	Value
1	Input Capacitor RMS ripple current	Cin IRMS	0.99 A
2	Output Capacitor RMS ripple current	Cout IRMS	0.19 A
3	Peak Current in IC for Steady State Operating Point	IC Ipk	2.3 A
4	Average input current	Iin Avg	0.93 A
5	Inductor ripple current, peak-to-peak value	L Ipp	0.65 A
2	Steady State PWM Duty Cycle, range limits from 0 to 100	Duty Cycle	44.6 %
3	Steady State Efficiency	Efficiency	89.3 %

$Io := 2 \quad \Delta I := 0.65$

$r := \dfrac{\Delta I}{Io} \quad r = 0.325$

$Cout\_Irms := \dfrac{r}{\sqrt{12}} \cdot Io$

$Cout\_Irms = 0.188$

$Vo := 5 \quad Vin := 12$

$\eta := 0.893$

$D := \dfrac{Vo}{Vin \cdot \eta}$

$D = 0.467$

2696

#	Description	Parameter	Value
1	Input Capacitor RMS ripple current	Cin IRMS	0.99 A
2	Output Capacitor RMS ripple current	Cout IRMS	0.17 A
3	Peak Current in IC for Steady State Operating Point	IC Ipk	2.3 A
4	Average input current	Iin Avg	0.93 A
5	Inductor ripple current, peak-to-peak value	L Ipp	0.57 A
1	Steady State PWM Duty Cycle, range limits from 0 to 100	Duty Cycle	44.8 %
2	Steady State Efficiency	Efficiency	89.5 %

$Io := 2 \quad \Delta I := 0.57$

$r := \dfrac{\Delta I}{Io} \quad r = 0.285$

$Cout\_Irms := \dfrac{r}{\sqrt{12}} \cdot Io$

$Cout\_Irms = 0.165$

$Vo := 5 \quad Vin := 12$

$\eta := 0.895$

$D := \dfrac{Vo}{Vin \cdot \eta}$

$D = 0.466$

2676

#	Description	Parameter	Value
1	Input Capacitor RMS ripple current	Cin IRMS	1.00 A
2	Output Capacitor RMS ripple current	Cout IRMS	0.07 A
3	Peak Current in IC for Steady State Operating Point	IC Ipk	2.3 A
4	ICs Maximum rated peak current	IC Ipk Max	3.9 A
5	Average input current	Iin Avg	1.00 A
6	Inductor ripple current, peak-to-peak value	L Ipp	0.55 A
2	Steady State PWM Duty Cycle, range limits from 0 to 100	Duty Cycle	46.4 %
3	Steady State Efficiency	Efficiency	90.1 %

$Io := 2 \quad \Delta I := 0.55$

$r := \dfrac{\Delta I}{Io} \quad r = 0.275$

$Cout\_Irms := \dfrac{r}{\sqrt{12}} \cdot Io$

$Cout\_Irms = 0.159$

$Vo := 5 \quad Vin := 12$

$\eta := 0.901$

$D := \dfrac{Vo}{Vin \cdot \eta}$

$D = 0.462$

2592HV

#	Description	Parameter	Value
1	Input Capacitor RMS ripple current	Cin IRMS	1.00 A
2	Output Capacitor RMS ripple current	Cout IRMS	0.12 A
3	Peak Current in IC for Steady State Operating Point	IC Ipk	2.2 A
4	Average input current	Iin Avg	1.00 A
5	Inductor ripple current, peak-to-peak value	L Ipp	0.40 A
2	Steady State PWM Duty Cycle, range limits from 0 to 100	Duty Cycle	47.7 %
3	Steady State Efficiency	Efficiency	83.4 %

$Io := 2 \quad \Delta I := 0.40$

$r := \dfrac{\Delta I}{Io} \quad r = 0.2$

$Cout\_Irms := \dfrac{r}{\sqrt{12}} \cdot Io$

$Cout\_Irms = 0.115$

$Vo := 5 \quad Vin := 12$

$\eta := 0.834$

$D := \dfrac{Vo}{Vin \cdot \eta}$

$D = 0.5$

Is the entire 267x family software missing a factor of about 2.3 here?

Figure 12-4 A Check of the Results Obtained from an Online Tool for the Buck Topology

IC, and the software tool may not have warned you. You should also observe that such software tools often do not consider important external component tolerances. For example, your inductor itself may have a ±20% tolerance on its inductance. But 20% lower inductance could mean up to 25% higher peak currents (if you are operating close to the CCM-DCM boundary, though with an *r* of 0.4, the increase in the peak would only be 0.2×0.25, i.e., 5%). Note that $1/0.8 = 1.25$, so if L falls 20%, the AC current increases 25%.

Question 6

January 2001 Hi, I am currently designing a DC-DC converter based on the 2577 with output voltages of 3.3V and 5V. I would like to know the methodology for selecting the

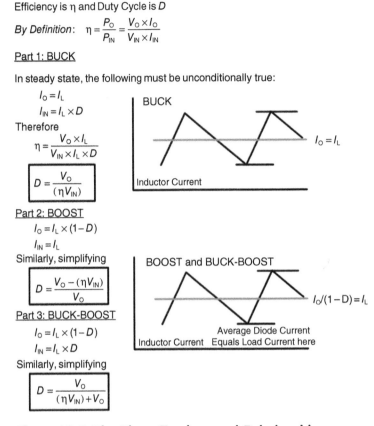

Efficiency is η and Duty Cycle is D

By Definition: $\eta = \dfrac{P_O}{P_{IN}} = \dfrac{V_O \times I_O}{V_{IN} \times I_{IN}}$

Part 1: BUCK

In steady state, the following must be unconditionally true:

$I_O = I_L$

$I_{IN} = I_L \times D$

Therefore

$\eta = \dfrac{V_O \times I_L}{V_{IN} \times I_L \times D}$

$$D = \frac{V_O}{(\eta V_{IN})}$$

BUCK

Inductor Current

$I_O = I_L$

Part 2: BOOST

$I_O = I_L \times (1 - D)$

$I_{IN} = I_L$

Similarly, simplifying

$$D = \frac{V_O - (\eta V_{IN})}{V_O}$$

Part 3: BUCK-BOOST

$I_O = I_L \times (1 - D)$

$I_{IN} = I_L \times D$

Similarly, simplifying

$$D = \frac{V_O}{(\eta V_{IN}) + V_O}$$

BOOST and BUCK-BOOST

Inductor Current   Average Diode Current
Equals Load Current here

$I_O / (1 - D) = I_L$

Figure 12-5 The Three Fundamental Relationships Connecting Actual Duty Cycle and Efficiency

voltage rating of the capacitor that will be used in the output filter. The capacitor value is 2200uF, but I want to know if I can use a 10V or 16V capacitor. These differ in their height, which really matters for my product. Also let me know how to go about derating and selecting the capacitor voltage for converters.

Blog Entry 1 Choose an output capacitor whose working voltage (WVDC) is at least 20% higher than the output voltage. Thus, the 10/16V capacitor you intend to use is overrated. A 6WVDC capacitor in this case should be sufficient for both outputs. There are no derating methods involved in choosing the capacitor's working voltage other than the selection criteria defined in the first sentence herewith. Other factors to consider besides WVDC are the ripple current and equivalent series resistance specifications. A detailed procedure for calculating these values is found in the datasheet of the 2577 on page 15.

Blog Entry 2   Hi, I presume that you intend to use aluminum electrolytic capacitors and not tantalum? I also presume you understand the effect that the ESR of the capacitor has on the output ripple voltage and how to determine what capacitor value is required? The ESR of a high-voltage aluminum electrolytic capacitor is usually quite a bit lower than the same capacitance value lower voltage rating capacitor. One would assume that a 3.3V supply shouldn't need anything higher than, say, a 4V rating and a 5V supply could use a 6V rating safely. But when the ESR of these are compared to 10V or 16V or even higher, the marked difference in ESR is very noticeable. However, this comes at an increase in can size and hence volume. One thing you need to be aware of is that aluminum electrolytics' ESR varies a lot with temperature. At room temperature and higher, the ESR is quite low, but at temperatures below about −10°C, the ESR climbs severely. Often if the equipment has to work over a wide temperature range, say −20°C to +60°C, the final choice of capacitor value and voltage rating will be determined by keeping the ESR sufficiently low at these low temps. This normally dictates the fitting of a much larger capacitance value than required by just the energy storage criteria. However often designers opt to use a slightly higher ESR capacitor. This is because the self-heating caused by the ripple current flowing in the ESR causes the capacitor to heat up enough to raise the internal temperature above the −10°C region. This isn't such a bad idea as the capacitor ESR falls dramatically at high temperatures, so high ambient temperature operation is not a problem. It also saves a bit on volume and costs. Hope this helps, Email me off-line if you need any further help.

End of Thread.

> *Author's Comments:* Nothing wrong here except that even the 20% derating guideline in the first entry may be too generous. For example, in AC-DC power supplies, we often prefer the cheaper 400V capacitors to the 450V capacitors, for the 385V HVDC rail of the front-end PFC stage. That actually leaves only 15V above 385V, that is, less than 4% derating. It is also true that many senior engineers disagree over this. However, it seems that today, there is no real statistical evidence suggesting that voltage derating enhances either the life or the reliability of most modern aluminum electrolytics. But you may need to check with your vendor.

> The second blog entry is from a senior defense engineer in South Africa. My impression is that he or she has very aptly expressed a lot of the finer aspects of selecting aluminum electrolytic capacitors. I suggest you read that very closely. It's folks like these (no hidden agendas, just helpfulness) that make it worthwhile.

> But wait a minute! The 2577 is a *Boost* switcher, requiring a minimum of 3.5V at its input. How is it that the first blogger (seemingly a company representative) did not even stop to ask how a Boost IC could be made to deliver 3.3V output from a 3.5V input? Or whether the input wasn't even lower!

Question 7

July 2001   We need to convert 9–14V DC to a 12V regulated output. Normally, one would use a flyback topology to achieve this. However, we would rather try to use a Boost topology with parallel 12V zener clamp diodes (for simplicity). The switcher simulations accept a Boost configuration with Vin of 9–12V. However, what would happen if we designed a circuit with the 2588-12 in Boost mode, but with a Vin of 9–14V? Regarding the zener diodes, should they then be placed between the Vin supply rails or between the Vout supply rails (before or after the feedback line)?

Blog Entry 1   I thought a little harder on how zeners work, and realized that placing the zeners between the Vin rails will cause the zeners to melt if >12V was constantly supplied with a large Vin source, since Vin is not current regulated. Please don't flame [blame] me for the zener question! I would still like to know how the Boost converter would react to overvoltaging. Perhaps that can answer whether zener diodes can even be used.

Blog Entry 2   I apologize for the long delay in my response to your inquiry. I skipped over your posting, thinking that someone must have replied to it already, only to find out that the entry logged against it was your own correction. In lieu of the zener diodes, you can use the 338 linear regulator to step down the +9V to +14V unregulated input voltage to +6V (the minimum Vin-to-Vout differential required by the 338 is 3V). Now, you can boost the +6V to +12V using the 2588-12. By doing this, the load current available will only be 2.25A due to the step-up conversion. The average input current to the 2588 will be slightly less than 5A. Thus, the 338 becomes suitable for this conversion because it is rated at 5A load current.

Blog Entry 3   A more simple solution to implement is to use the 3478 Low-side N-channel controller in a SEPIC configuration. This application utilizes two inductors (instead of a Flyback transformer) to attain the Buck-Boost function. The 3478 requires an external user-selectable Fet switch, so you can choose the one that suits your load current requirement. The datasheet provides an application rationale for SEPIC configuration on page 19, Figure 13. The output voltage can be set to 12V by changing the value of the feedback resistor.

Blog Entry 4   The best way to do this is to put in a low dropout linear regulator on the output. Set the Boost regulator output for 12.5–12.7V, and then use an LDO to drop your voltage to 12V.

End of Thread.

> *Author's Comments:* The first official suggestion (without even asking what the required load current is) is to use a 5A linear regulator to drop the 9–14V to 6V. Try dropping 14V to 6V at 5A. The dissipation is $(14 - 6) \times 5 = 45\text{W}$. The junction-to-ambient thermal resistance of the package is 35°C/W, so it is really lucky the linear regulator has overtemperature protection! Assuming it can deliver 5A to the 2588 (however momentarily), that DC-DC stage will try to step it up from 6V to 12V. Since its duty cycle is about 50%, the average inductor current will be $I_O/(1 - D) = I_{IN}$, and therefore $I_O \approx 2.5\text{A}$. The representative has correctly figured out the maximum load current, except for the fact that the dissipation in the linear regulator will be too high.

The second person, who judging by the tone is possibly their Applications Manager, has given the "best" advice in the matter. Blogger 2's solution will work, except for the fact that he or she has not explained how he or she intends a Boost working at 14V to deliver a regulated 12.7V!

In general, remember that doing anything at the *input of a Buck* is usually preferable to trying to achieve the same function at its output, since the current at the input of a Buck is lower than at its output. Similarly, for a *Boost*, you would prefer to put a switch or pass element *at its output*, not at its input, for the very same reason. For example, recently I suggested that the load disconnect Fet we wanted to introduce in our new Boost regulator IC be placed at the output rather than the output. There are, however, pros and cons of doing that. As our senior IC designer correctly pointed out, if the Fet were placed at the output, it would actually involve using the company's high-voltage transistors (which were about 3× larger in area), whereas at the input, they could still use their low-voltage transistors. Knowing that the high-voltage transistors took in a lot of die area, we could even hope to achieve better efficiency in a smaller die area using lower-Rds Fets placed at the input rather than the output of this Boost.

Always *think everything through*. Play the Devil's Advocate. *Try to prove yourself wrong*, not always right (do it silently in your mind if that is more comforting to you). You will be surprised how often *you will lose to yourself*! Better than losing to others, you might say! As I said before, in Power Conversion, the most obvious conclusion is often the wrong one. And I was clearly the one wrong above. So I also learned—*Apps and Design need to work together to come up with the best solutions*. Neither of them can do well in isolation.

There are other ways to achieve a non-inverting step-up/step-down function that should also be evaluated. One is to use a Flyback topology with a transformer to correct the natural polarity inversion of the Buck-Boost topology. Another way is to use the "4-switch Buck-Boost." You can refer to my application note AN-3247 at www.freescale.com for a detailed understanding of this topology. What few people realize is that you can come up with a simple "brute-force" 4-switch Buck-Boost without adding any additional control or frills. Just take a standard N-Fet based Non-Synchronous Buck switcher, and add one additional diode and one external Fet, as shown in Figure 12-6. The efficiency will not be very good (around 60 to 70% typically), but it is worth considering for its sheer simplicity. Note that the Gate of the second Fet can usually be directly attached to the SW node (the Source of the upper Fet). Just watch out that you don't exceed the Gate-to-Source ratings of the lower Fet. The two Fets turn ON together, delivering energy into the inductor. Then both turn OFF together and the current freewheels into the output. Basically, the additional diode and Fet constitute the price you are paying for correcting the traditional polarity inversion of the conventional Buck-Boost topology.

Lest we forget, the customer also asked another question that remained unanswered until the end—what if the customer set up a Boost for a 12V output and the input goes up, say, to 14V. Well, the regulator would run open loop, and the output would stay roughly 500mV below the input rail. The Fet would turn OFF completely sensing the high voltage on its feedback pin, with the Schottky diode just dropping the output in series with the input. So now the customer could put in an LDO at the output set to deliver 12V from the $14 - 0.5 = 13.5$V rail coming in.

One more thing! Do a search on any such discussion forum for *all questions* relating to any part you want to use. For example, look under "3478" (or its family), under discussion here. You are likely to find the maximum number of queries for this particular part on the forum. Why so? Remember, this is a Flyback controller with a *100% duty cycle*. But don't expect the company to spell that out to you!

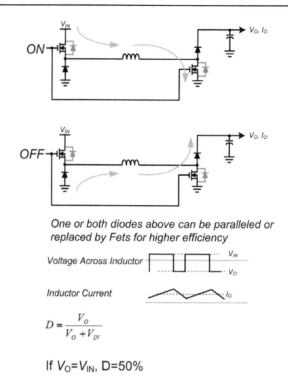

One or both diodes above can be paralleled or
replaced by Fets for higher efficiency

$$D = \frac{V_o}{V_o + V_{IN}}$$

If $V_O = V_{IN}$, D=50%

**Figure 12-6 A Simple 4-switch Non-
inverting Buck-Boost**

What follows is what you should really know about the Boost and Buck-Boost topologies themselves.

One of the common aspects of these two topologies is that in both of these, energy is built up in the inductor during the switch on-time, during which duration, *none* passes to the output. Energy is *delivered to the load only when the switch turns OFF*. In other words, we have to turn the switch OFF to get any energy at all delivered to the output. Contrast this with a Buck, in which the inductor, being in series with the load, delivers energy to the load even as it is being built up in the inductor itself (during the switch on-time). So in a Buck, even if we have 100% duty cycle (i.e., switch ON for a long time), we *will* get the output voltage to rise (smoothly). Subsequently, the feedback loop will command the duty cycle to decrease when the required output voltage is reached.

However, in the Boost and Buck-Boost topologies, if we keep the switch ON permanently, "because the output is low," we can actually *never* get the output to rise. Remember, in these topologies, energy is delivered to the output *only* when the switch turns OFF. We can thus easily get into a Catch-22 situation, where the controller thinks it is not doing enough to get the output to rise—and therefore continues to command maximum duty cycle. But with a maximum 100% duty cycle, that means *zero* off-time—so how can the output *ever* rise? We can get trapped in this illogical mode for a long time, and the switch can be destroyed. Of course, we are *hoping* that the current limit circuit is designed well enough to eventually intervene, and turn the switch OFF before the switch destructs! But you will be applying a lot of stresses on the switch every time you power up. So, generally, it is considered inadvisable to run

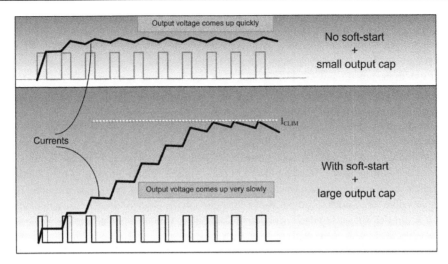

Figure 12-7 Soft-start May Do Nothing to Control the Peak Currents During Power-up

either of these two topologies at 100% duty cycle. It is just *illogical*. And sooner or later that oversight will return to bite you.

Another myth being constantly propagated on such forums is that *soft-start will always help reduce the stresses in the switch at turn-on*. *It does not*, for the simple reason that the current and the duty cycle have no simple relationship in CCM. So a 5V to 3.3V switcher will have a duty cycle of about 66% if the load current is 2A, it will have almost the same duty cycle if the load current is 4A, and so on. So how can you ever limit the current based on duty cycle? You don't. Take a look at Figure 12-7. The peak current mainly depends on the amount of output bulk capacitance. This is because it is only when the output capacitor voltage comes up does it slowly decrease the up-slope of the inductor current (i.e., $V_{IN} - V_O/L$) and increase the down-slope (i.e., V_O/L), and thereby reach a certain balance. Until that happens, a steady state cannot be achieved, because after all, voltseconds balance needs to finally occur. So the current keeps ramping up steeply, barely ramping down, and the duty cycle can do almost nothing to forestall that process. Yes, current limit can ultimately enter the picture, but its main job is to save the switch, nothing more.

Many switcher ICs are in fact designed with a certain minimum on-time (especially the current mode control types). They also keep to the minimum pulse width until about 0.2 to 0.3V on the feedback pin. In such cases, with a reasonably large output bulk capacitor, *you will see a huge inrush of current into the output capacitor, even before the latter starts to rise appreciably*. You should also be aware that inrush current into the input capacitor of any topology is very high, and no switch action can even hope to prevent that.

So where does soft-start really help? Mainly in bringing up the output voltage rail smoothly, maybe to avoid jerking the system connected at the output of our power supply. Yes, perhaps within a *certain range* of output C and L, it can also help control the stresses on the switch at power-up. But that support is hardly unconditional. Maybe it helps somewhat in lowering the overshoot of the rail at startup. But as mentioned in Question 2, *it can itself be a reason for the overshoot too*. Therefore, in all the AC-DC Flybacks we designed in Singapore, not one ever had any deliberate in-built soft-start. Why would we do that, if the customer has not explicitly requested it?

An exception is the Boost converter. When you power that up, there is a huge inrush current into the *output* capacitor too, in the initial moments, even though your switch may be completely OFF. Unfortunately if the switch tries to come ON at that very moment, it diverts this huge current into itself. This can kill the switch, because though it may even be starting up with a very small duty cycle (either the result of soft-start, or because it hit current limit), as we can see from Figure 12-7, the current can keep ramping up in the switch with almost no control. So yes, in a Boost, soft-start *can* help you save the switch. But my personal preference is not a soft-start, but a time delay—where you wait a little for the inrush to be over before you even start switching. In high-power PFC Boost stages, it is common to put an additional diode with its anode at the input rail and its cathode at the positive terminal of the output capacitor. The inrush therefore gets diverted through this diode rather than through the inductor and Boost catch diode. So even if the switch turns ON, basically there is no current through it. Once the inrush is complete, and the output starts rising higher than the input, the extra diode gets permanently reverse-biased and goes out of the picture automatically.

Question 8

June 2001   I'm using the 2677 to generate a 15V/4A rail from 24V (Buck regulator, standard topology). If I let it regulate with no load and then switch in a 4A load, I see approximately a 10V instantaneous droop in the output. The output recovers to 85–95% of regulation within a few ms, but the last portion of recovery can take seconds to minutes and seems to vary, based on the circuit's temperature. The droop isn't a big deal but the time response to regulation is. The only solution I've seemed to find that helps is to add a 200pF cap between the switch output and ground. The regulator then seems to work much better getting back to nominal regulation within 2ms. I'd like help on what is happening and why the cap fixes it, so I can maybe get the right components in the next switcher I'm doing.

Blog Entry 1   Please realize that I do not have enough information to give a definitive answer. However, based on the information you have provided I would guess that the problem is noise-related. By this, I mean noise is being generated, which is being injected into the control loop or internal circuitry of the 2677. Reverse recovery in the diode could be causing some current spikes which result in noise. To reduce this a snubber may be used. See the 2679 Eval board App note. Also, for high current applications, a 1uF ceramic right at the Vin pin and ground helps provide some bypassing for the high-frequency components of the switching current (Cinx as listed in the App note). A 10–100 Ohm resistor can also be placed in series with Cboot to reduce the turn-on time of the main power Fet, which results in lower MHz level noise. Efficiency loss is minimal. Lastly, we have found some diodes have a LOT more capacitance than others. Just by changing the diode we have solved noise and circuit problems. Hope this helps. The 267x devices typically work well, but the higher currents do result in more problems.

Blog Entry 2 You are right. I did not have any answer to my problem of load transient while using 267xT-Adj device—until I saw this reply. Because I simply followed your software tool, which has no mention about Cinx. But after introducing Cinx the output is stable under all load conditions. Thanks.

End of Thread.

Author's Comments: Historically, this was probably the earliest sign of the looming $D > 0.5$ problem with the 267x family. Luckily in this case D was only slightly higher than 50%, *though only with the input set at its nominal value.* So even at this moment, if the company had shown due diligence and marshaled their Apps resources to actually set up a board with $V_O = 15V$ (which they obviously had never done before), they would have caught the problem. Instead the first responder was busy brushing off this inquiry with "I do not have enough information" (well then get it if you are the Apps Manager). It took them about 1–1/2 years (and a very irate and persistent subsequent customer) to finally realize they had a serious problem on hand. Note also that the input decoupling capacitor ("Cinx") is still missing from their latest software release (v. 6.24)—that's six years after this complaint was lodged. Even up to the present day, the referred eval board App note still shows no component marked "Cinx" on the PCB layout. Actually, all the four input capacitors are marked "Cin," so you may need to be fairly smart to figure out which one was *supposed* to be Cinx.

The importance of good input decoupling has been discussed in great detail in Chapter 2. The preferred value of the capacitor is actually 0.1μF, because it has a 30MHz bandwidth, unlike a 1μF capacitor that only goes up to about 10MHz. The resistor in series with the bootstrap capacitor has also been discussed in Chapter 10. Though you will notice that this suggested location of bootstrap resistor actually helps the *IC* to work better too—not merely by slowing down the turn-on transition. In fact the former explanation seems more applicable here, as it is consistent with the observation that "Efficiency loss is minimal." This could be picked up by a very perceptive engineer as an indication that all is not well with this IC, and the IC may be excessively noise-prone. It is far easier to figure out the rather casual appreciation of customers' time and money.

A clue provided by the customer is that a ceramic cap from SW to ground helps. So yes, *noise is also an issue here*. Hearing this, I would have immediately suggested a PCB layout review, and if necessary a small RC snubber (10Ω to 100Ω in series with a 470pF to 4.7nF capacitor) between SW node and ground. The advantage of an RC snubber instead of just a "C" is that the energy in the capacitor is dumped in the "R" of the RC snubber, not in the switch.

Incidentally, Schottky diodes do *not* have any reverse recovery current issues, but they do have some body capacitance, which can produce similar effects. However, that fear has always proven to be exaggerated. I have personally not seen any application having a performance issue explicitly related to a "bad Schottky." The only exception was a case where the leakage current of the Schottky was so high, it was prematurely tripping the current comparators inside the switcher IC. And there was also a *reliability* issue once, concerning the dV/dt rating of a commercial Schottky. Both these issues are discussed elsewhere in this book.

Question 9

September 2001 I am using the 2621 to step up 3.6V to 5V at 300mA. C1 and C2 are both 47uF Tantalum capacitors. The diode is 21DQ04. The inductance is 6.8uH, as recommended in the 2621 datasheet. It is the same part number as recommended in the datasheet. Also the PCB layout is according to the datasheet. But the 5V output ripple is high (500mV). What can I do to reduce the output ripple?

Blog Entry 1   If the objectionable ripple occurs at the switching frequency (i.e., you measure a 200kHz component as per the App Note circuit), then make sure that the negative terminals of C1, C2, Pin 1, and the output return terminal are located close together. The output should be taken from the terminals of C2. Don't use more than one measurement grounding point. Some high-frequency noise (10 to 25MHz) may be generated in the diode when it is switched off. A small bead on the diode lead may reduce it. If the objectionable ripple occurs at a subharmonic of the switching frequency, or at the hysteretic period (light load), then check that the timing capacitor and the 0.1uF VDD decoupler are connected directly to the signal return pin. Recheck the size of Cf1.

Blog Entry 2   Thanks for your reply. I tried placing the components in every possible way. But try and try, I just can't make C1, C2, Pin 1, and GND all close together, along with the diode also close to pin 8. I [just] can't do it. So please send me a demo PCB picture. Thanks.

Blog Entry 3   Attached is the schematic of the 2621 design you requested, including the list of recommended parts to build it. I'll try to send you the recommended layout of the printed circuit board under a separate response as soon as I find out how to do it.

End of Thread.

> *Author's Comments:* Many PCB layout guides seem to think it is physically possible to put *every* component right next to the IC (overlapping?). Without any help from your Physics professor on how to increase the number of spatial dimensions, your best chance here is to create a *priority list*, that is, which component takes precedence over the other. Layout has been discussed in great detail in previous chapters, and it should be consulted. Though, unfortunately, hysteretic ICs such as the 2621 are even harder to design in a layout, and still are often unpredictable.

> We know from Chapter 2, that a schematic provides very few clues, if at all. I started wondering if this customer ever got the promised "demo PCB picture." That took me to the datasheet of the 2621 on the web. Here's what it says about the recommended PCB layout:

>> High switching frequencies and high peak currents make a proper layout of the PC board an important part of design. Poor design can cause excessive EMI and ground-bounce, both of which can cause malfunction and loss of regulation by corrupting voltage feedback signal and injecting noise into the control section. *Power components—such as the inductor, input and output filter capacitors, and output diode—should be placed as close to the regulator IC as possible, and their traces should be kept short, direct, and wide. The ground pins of the input and output filter capacitors and the PGND and SGND pins of the 2621 should be connected using short, direct,

and wide traces. *The voltage feedback network* (Rf1, Rf2, and Cf1) *should be kept very close to the FB pin.*

Considering that the layout was considered so important (and rightly so), the datasheet still carries no pictures of a PCB layout. There also seems to be no accompanying App Note with that vital information. Though there is an eval board available from the company's website that you have to cough up $30 for. Short of wanting to buy it just to confirm the layout, their web site left me a little confused, but my initial impression was that the 2621 eval board features only the Sepic topology (not what this customer wanted, i.e., a Boost).

Question 10

September 2001 Dear Sir, we are using the 2575-5.0 for deriving a 5V output for an input voltage varying from 8 to 13V.The load is 0.8A. We have the following queries in regard to this design. The value of the inductor is specified as 330uH on page 1 of the datasheet (for an input voltage from 7–40V at 1A). But there is an inductor selection curve in the datasheet (Figure 4), which recommends 220uH for our specified voltage and current range. But your downloadable software shows the recommended value of 330uH. Kindly suggest what value of inductor is applicable for our design. Also, the noise level required for our application needs to be less than 50mV. Your software shows the ripple level as being greater than 60mV. But in the datasheet performance curve (page 9), the ripple level is shown as 20mV. Practically we have observed the ripple voltage to be around 45 to 50mV. Let us know the applicable noise limits for this switching regulator.

Blog Entry 1 The value given in the datasheet generally covers a wide range of load current spec. The following lists the recommended components for your specific application—CB: 0.01uF; Cin: 80uF; Cout: 68uF; D1: General Semiconductor SS24L1, Inductor: 47uH. All of the above values were obtained using our online virtual bench tool for switcher design. Please try it, it really helps.

End of Thread.

> *Author's Comments:* Let us do a quick sanity check here. According to the company official, the right value is 47μH. The duty cycle is about 0.5. The switching frequency of this part is 52kHz, so the time period is about 20μs. Therefore the off-time is about 10μs. If the output is 5V, the ΔI is $V\Delta t/L = 5 \times 10/47 \approx 1A$. This is way too much. You typically want a current ripple of ±20%, that is, a ΔI of $0.4 \times I_0 = 0.32A$ in our case. To decrease the ΔI from 1A to the more appropriate 0.32A, we need to actually increase the inductance three times, that is, from 47μH to about 150μH. And that is closer to what the datasheet says. So in this case the advice to move away from the datasheet to the software tool is actually strange.
>
> As for the customer's observation that the datasheet claims the ripple will be ±20mV, you should be aware that almost all companies deliberately choose the best conditions to present their curves. Remember, these curves are only "typical," and your definition of what that word means can differ dramatically from theirs. From their point of view, it is a dog-eat-dog world out there, and if they don't do this, the customer will likely penalize them and pick the competitor's parts without further thought. They also figure that very few

customers penalize anyone for *slowly-discovered specmanship*. They probably have almost nothing to lose, and everything to gain by continuing the tradition. Specmanship is almost considered acceptable today, so why would they ever try and *buck* the trend? But why boost it, pray? At least maintain regulation.

But in the case of the 2575, I wanted to double-check it against their current software, hoping they had corrected it by now. I found that though the inductor automatically chosen was now 330μH, no operating values were provided, and further, it carried a disclaimer that the values of the schematic are *not* optimized for any specific application (then why go through the effort of putting out a tool anyway?). So I decided to check out their latest generation switcher device (with emulated current mode control or "ECM"), the 25575, which was actually proffered as the very first choice by their software for my proposed application.

The results of that were tallied against some simple Mathcad checks as shown in Figure 12-8. Once again, neither have they got the output cap RMS formula right (after probably a *decade* in the business of writing software for their products), nor did they even understand the simple universality of the Buck topology equation $D = V_O/(V_{IN} \times \text{Efficiency})$ given in Figure 12-5. Also, the software display panel states the duty cycle range is "0 to 100%." It really isn't, because this part has a whopping fixed off-time of 500ns (to help it sample the diode waveform to generate its emulated ramp). Knowing it has such a big off-time (i.e., a small D_{MAX} in effect), I found it surprising that this part is supposedly adjustable up to 1MHz. Because in that case, the time period is 1μs. And a 500ns off-time means a D_{MAX} limit of about 50%. In other words, if the input is 8V, your maximum output is only about 3.5V (allowing about 0.5V for the diode). Luckily, for the 5V/0.8A application I chose, the software correctly chose a maximum switching frequency of 390kHz. So are you telling me it is *not* adjustable to 1MHz, huh?

Incidentally, the famous Robert Pease book, *Troubleshooting Analog Circuits* carries an Appendix F titled "How to Get the Right Information from a Data Sheet." This was also repeated as Appendix F in every linear databook his company published (haven't seen those for a long time now, though). The legendary author rightly crusades against misleading information in datasheets. He warns you about not believing anything that is labeled a "typical," and/or is not a "guaranteed spec (i.e., MIN and MAX)." I agree, you should be very careful with your expectations about a part. Also, read the actual *test conditions* used for the stated parameter within the Electrical Characteristics Tables.

With all that in mind I started looking at the 25575 datasheet in detail. The only mention of the *actual effect* on the dropout voltage (or equivalently on the minimum V_{IN} or the maximum frequency achievable) due to the massive 500ns off-time, was tucked deep into page 13 of the datasheet. Not a word on the cover page, not anything in the Tables (not even a fine print "Note" at the end of it), nothing in the Typical Performance curves either. Also note that they guarantee the frequency in the Electrical Tables in the following way. They just pick one or two arbitrary frequencies, provide a spread for them corresponding to a certain *RT* (the frequency adjust resistor). So if you are using any other resistor, you will have to look at their typical performance curves. But those aren't guaranteed! You really don't have a single card in hand.

I wanted to derive the self-confessed "approximate" equation on page 13 of their datasheet. This derivation is in Figure 12-9 and Figure 12-10. I also plugged the resulting equation into Mathcad, as shown in Figure 12-11, to get a real feel for what was happening. The equations derived in Figure 12-9 are actually valid for any Buck converter. They tell you that, if, for example, you are trying to step-down 45V to 1V, you may run into a D_{MIN} issue. Your converter may not be able to reduce its duty cycle sufficiently because of a minimum on-time t_{ONMIN}. This is also related to the forward drops across the switch and diode. Similarly, as in the case of the 25575, if you are hoping to go, say, from 7V to 5V, that may not be possible if your converter has a D_{MAX} limit and/or the switch and diode drops are excessive. Using these equations I finally

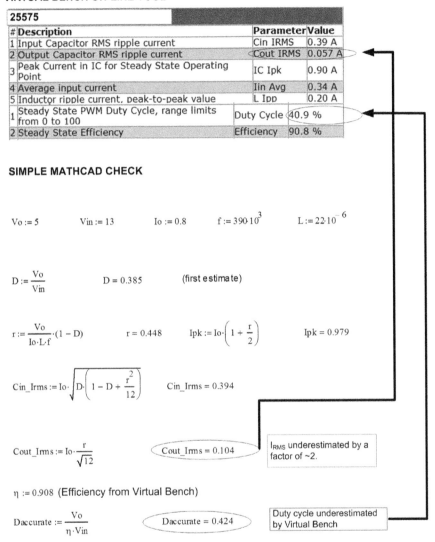

Figure 12-8 A Simple Check of the Results Provided by an Online Tool for an "Emulated Current Mode Control" Switcher

derived the said page 13 equation at the end of Figure 12-10. I realized they had ignored the switch drop entirely, and also part of the diode drop to get there. But the switch Rds is as high as 0.66Ω, which leads to a drop of 0.99V at the maximum rated 1.5A of the device. That is not insignificant, I would have thought.

But there is finally some good news too! Looking at the Mathcad spreadsheet in Figure 12-11, I realized that for the original customer requirement of 5V/0.8A from an 8–13V input, the software actually picked a

Buck Duty Cycle (assuming CCM):

$$D = \frac{V_O + V_D}{V_{IN} - V_{SW} + V_D}$$

Therefore

$$V_{IN} = \frac{1}{D} \times (V_O + V_D) + V_{SW} - V_D$$

$$V_{IN} = \frac{V_O}{D} + \frac{1-D}{D} \times V_D + V_{SW}$$

For a given output voltage

$$V_{INMAX} = \frac{V_O}{D_{MIN}} + \frac{1-D_{MIN}}{D_{MIN}} \times V_D + V_{SW}$$

where $D_{MIN} = t_{ONMIN} \times f$

Similarly

$$V_{INMIN} = \frac{V_O}{D_{MAX}} + \frac{1-D_{MAX}}{D_{MAX}} \times V_D + V_{SW}$$

where

$$D_{MAX} = t_{ONMAX} \times f = (T - t_{OFFMIN}) \times f = 1 - t_{OFFMIN} \times f$$

Figure 12-9 Minimum and Maximum and Minimum Input Voltages Possible for a Buck, Based on Minimum On-time and Minimum Off-time (Or Corresponding Duty Cycles)

relatively safe frequency of 390kHz in Figure 12-8. To the right of the plot "Dropout vs. frequency," you see that that frequency will suffice down to an input of 6.86V, about 1.25V lower than the minimum requirement of 8V in the application. So they do have healthy amounts of "guardbands" or "fudge-factors" in the software to hopefully take care of the lack of important guaranteed specs in the datasheet. All the same, I think the guys who wrote the datasheet certainly need to visit (or revisit) "Appendix F." Because the plot in Figure 12-11 also tells me that at 1MHz, the dropout is about 6V. That means, to get a 5V output, you need to stay above 11V at the input (while marveling at the benefits of "ECM").

As you can see, it is the minimum off-time of 500ns that can be devastating for emulated current mode control ICs. It ends up forcing you to decrease the frequency significantly if you ever want V_{IN} to even approach V_O (and in modern applications that is increasingly important). Apparently conscious of that frequency limiting issue, the company succeeded in diverting attention by focusing on the acceptable minimum on-time of the IC, rather than the minimum off-time—see Figure 12-12, reproduced almost exactly from their website. Their graph is actually very obvious—they just picked a very *high* V_{IN} of 36V to show how *low* they can go from there, based on their 80ns minimum on-time. They also seem to be comparing it with some "competitor" who is apparently still struggling with a fairly bad 150ns of minimum on-time, thereby also implying they can actually go to a much higher frequency than a "2.8MHz" switcher! Interesting marketing, really! However, as an engineer, you cannot afford to forget the ramifications of

Minimum Dropout Calculation for ECMICs

We have from previous figure (*assuming* $V_{sw} = 0$):

$$V_{INMIN} = \frac{V_O}{D_{MAX}} + \frac{1 - D_{MAX}}{D_{MAX}} \times V_D$$

Since minimum off-time for ECM ICs is 500 ns currently,

$$D_{MAX} = 1 - t_{OFFMIN} \times f = 1 - 500\,ns \times f$$

Therefore

$$V_{INMIN} = \frac{V_O + V_D - V_D \times (1 - 500\,ns \times f)}{(1 - 500\,ns \times f)}$$

$$V_{INMIN} = \frac{V_O + V_D}{(1 - 500\,ns \times f)}$$

(assuming $V_D \times (1 - 500\,ns \times f)$ is negligible!)

This is the equation given in the datasheet, but with several assumptions.

Figure 12-10 Minimum Input Voltage for the 25575 Emulated Current Mode Control IC

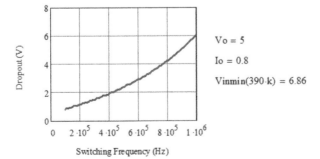

$$k := 10^3$$

$$Vo := 5 \qquad Io := 0.8 \qquad Rds := 0.66 \qquad Vd := 0.5$$

$$f := 100 \cdot k, 110 \cdot k .. 1000 \cdot k$$

$$Dmax(f) := 1 - 500 \cdot 10^{-9} \cdot f \qquad \text{(assuming minimum off-time 500\,ns}$$

$$Vinmin(f) := \frac{Vo}{Dmax(f)} + \frac{1 - Dmax(f)}{Dmax(f)} \cdot Vd + Io \cdot Rds$$

$$Vo = 5$$

$$Io = 0.8$$

$$Vinmin(390 \cdot k) = 6.86$$

Figure 12-11 Mathcad Verification of the Minimum Input Voltage for the 25575 Emulated Current Mode Control IC

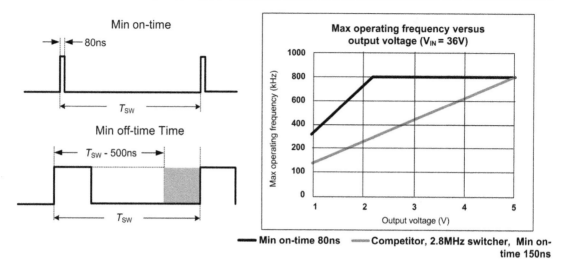

Figure 12-12 A Page from a January 2007 Media Presentation for Emulated Current Mode Control ICs

either the minimum on-time or the minimum off-time. What really matters to you depends on your exact application.

Question 11

August 2006   Great part guys. I am having great success with this part in a new design, which is an interleaved, single-output converter. The one sticking point I have is that at approx 50% duty cycle, there is a lot of PWM jitter. I have tried all combinations of placement of current sense bypass caps (at the IC end) and current sense components that I can think of. Likewise, I have studied both the App Note and the IC datasheet. Perhaps there is an issue with slope compensation. Unfortunately, there is no way to add the clock signal to current sense (the classic way to introduce slope compensation), because the 5034 has a DC level controlled clock pin. Any suggestions would be greatly appreciated! Thanks.

Blog Entry 1   I have used many 2-phase chips from many vendors and what you are seeing is typical. Basically if phase A is running near 50%, the noise caused by phase B turning on 180 degrees out of phase gets into the error amp output or into the ramp generator. Thus, you get jitter. The only way the IC vendor can avoid this is by using two separate die in one package. Even then, noise can find its way through the feedback pin. You need to find out if they have any tricks to get around the problem. Looking at the App Note for the 5034 Eval board, the author is the guy who knows the answer to your question.

End of Thread.

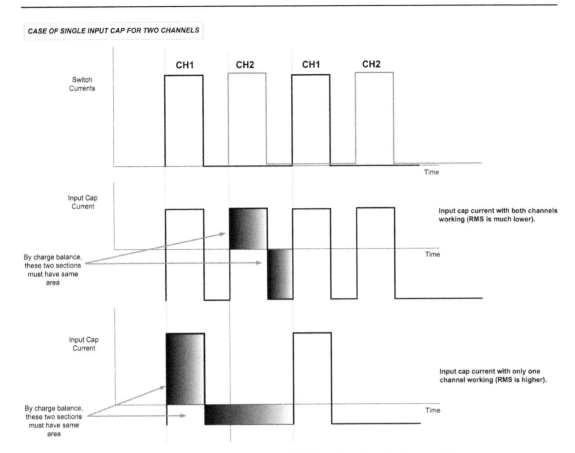

Figure 12-13 Input Capacitor Current Possibilities for Dual Channel Out-of-phase Buck Switchers

Author's Comments: The "classic way" the customer is referring to is described in that brief exchange with the "overheated Boss" in Chapter 1. The helpful reply (apparently proffered by some unconnected engineer) is not only encouraging, but in fact correct. However there is more to add. Many vendors are nowadays making *independent* dual-channel ("interleaved") regulator ICs with the promise that running them 180° apart reduces the RMS current (and stress) on the input capacitor. The assumption is that there is only *one input capacitor* to handle both the channels. So they also show typical schematics or marketing documents, highlighting this "advantage." The situation is shown in Figure 12-13. Note firstly that in calculating the worst-case RMS current, the situation can actually get worse if one channel is disabled (or running lightly loaded). Because the reduction in RMS actually depends on *both* channels being loaded to the max and thereby "balancing" the input capacitor waveform better around the 0A axis. If that does not happen, the input capacitor waveform can get skewed to one side of the 0A axis, thereby *increasing* the RMS current. That happens because the capacitor current waveform must have equal areas above and below the 0A axis (by charge balance) no matter what. Note that no RMS calculation ever depends on the frequency, it only depends on the basic "shape." So, in the end, the "advantage" of independent-channel interleaved independent converters may be, at best, tenuous. There is another issue that comes to light if you look at their eval boards carefully, and compare it to their "typical applications schematics." On the eval boards, you will almost invariably find that vendors of such ICs have in fact provided *separate* input capacitors for both channels. Because, as the blogger pointed out above, the two channels can end up feeding plenty of noise into each other on the output rails. In particular, if the duty cycle of both converters is close to 50%, the noise generated by Channel A turning ON just around the point where Channel B was *about* to turn OFF often injects noise onto the PWM comparator inputs of Channel B, causing a premature termination of its on-pulse. This leads to significant observed jitter. Yes, separate die would probably help, but it would be an extreme solution. The bare minimum you can do is to separate the power stages as much as possible—by using separate input capacitors to start with!

But if you have two or more switching channels being used to generate *a single output rail*, that is truly an "interleaved" supply, and most of the arguments in favor of interleaving for input capacitor reduction do apply.

Question 12

July 2006   I would like to know how to calculate the minimum load current which an SMPS can deliver, in order that its regulation is still OK. Indeed, I would like to use the 2594HV for my design. This power supply is always going to be active. What happens when the different components (loads, etc.) connected to this regulator's output are in standby mode and only consume about 100uA? Will the regulator still work? What is the consumption of the chip in such a mode? Thank you.

Blog Entry 1   See the attached file for the answer to your request.

Blog Entry 2   First, the 2594HV will regulate just fine with no load connected to the output. The feedback resistor load is sufficient. You may find the part will skip cycles, and this is normal for many switching regulators at no load. The regulator will still consume about 5mA at no load. If you need less, consider the 5007 (0.5mA). If you have a high-voltage input, I recommend the 2597HV where you can run the chip from the output voltage, provided it is set between 3.5V and 30V. What the previous person gave you is the equation that determines when you move from continuous conduction mode to discontinuous conduction mode. Discontinuous mode refers to the inductor current falling to 0.0A and staying there. If one used a Synchronous Buck regulator, you would be in continuous mode even at no load because the current in the inductor would reverse. The only reason "continuous" vs. "discontinuous" is of any concern is because the control loop changes characteristics. In general, when moving from continuous mode to discontinuous mode, the bandwidth reduces, but the stability improves. I have almost never had a customer who found a system problem as a result.

End of Thread.

> **Author's Comments:** About 5 years later, they apparently had the same (incorrect/inadequate/irrelevant) stock reply to give as in Question 3. It seems their Apps manager probably stepped in to save the show. His

or her answer, despite being almost casually delivered, is mostly correct, at least technically speaking. I did find some surprising "typos" and unconventional phrases that I had to correct in his or her original statement on the web. In particular, one of his or her original statements was (quoted verbatim) "continuous mode condiction refers to the inductor current falling to 0.0A and staying there." I hope he or she really meant "discontinuous conduction mode." As you can see, I did provide benefit of doubt, despite the fact that the original phrase had in effect, not one, but *three* typos in as many words. We are all just too busy sometimes, and that can attract plenty of errors. Typos are the least of all the magnificent mistakes we can make.

Question 13

July 2006 I am trying to design a simple switcher using a 2674 that supplies a post-regulator for 3.3V and other 5V devices. The entire device is battery powered and needs high efficiency. I don't want to switch off the regulator, and it will be on always. But the CPU goes into power-save mode. In this condition, the current to the circuit will be less than 50uA. My question is—is the 2674 suitable for this application (specifically in power-save mode with low current consumption)? Secondly, how can I calculate the efficiency for this mode?

Blog Entry 1 The quiescent current of the 2674 when it is in shutdown mode, i.e., when it is not switching and therefore not supplying power to the load is 50uA. The efficiency of the switcher during this mode is 0%, as given by: Eff = Pout/Pin. Assuming Vin = 24V, if Pout is 0.0W and Pin = 50uA × 24V = 1.2mW, then Eff = 0/1.2mW = 0.

Blog Entry 2 Check out the 26001. It is state-of-the-art. It only uses 40uA when not switching, but maintaining output regulation. It is a 1.5A part, but is far superior to the 2674 in your application. The 2674 uses 2.5mA when connected to Vin. If the load is 50uA, the efficiency will be very poor. A trade-off device might be the 25007 or the 2694. They both use about 500uA at no load.

End of Thread.

> *Author's Comments:* The customer is asking a very simple question—he or she will *not* be switching OFF the system, and so the minimum load on the converter could be as low as 50μA, but the rail has to be up all the time. The first reply talks about shutdown mode, that is, when the IC is internally powered OFF completely. So it is neither switching, nor is there an output rail. Not exactly what the customer asked for! The second company official said "in your application," and then tried to direct the customer away to some other parts. But the customer has *not* really revealed anything much about his or her application yet. He or she should have been asked. All he or she had said is the equipment is "battery powered," and that he or she needs 3.3V and 5V post-regulators. But it is not at all clear what the input range or load requirement is. The fact that the customer is planning on post-regulators of 3.3V and 5V implies that the input is not the usual single-Li-ion or 2–3 AA/AAA battery packs. Maybe a 9V pack. Maybe 12V. So the company official may well be trying to sell a comb to a bald man. In fact, 500μA at no load is usually considered extraordinarily high for portable battery-powered equipment—yes, that would be some tradeoff for sure. I just hope the second blogger didn't happen to be the company's Apps manager recently promoted to Marketing Director position of the product line. Because Marketing should really mean more than just that!

Also look at the rather extraordinary datasheet of the 26001. This is what caught my eye:

a) On the front page, under "General Description," it states that "the part has a wide input voltage range of 4V to 38V and can operate with input voltages as low as 3V *during line transients*."

b) On the front page, under "Features," it states "4.0V to 38V *continuous* input range."

c) Above the Electrical Characteristic Tables, under "Operating Ratings," it says "Supply Voltage (Note 4): 3.0V to 38V."

d) Under the fine print called "Note 4," it says "Below 4.0V input, power dissipation may increase due to increased Rds(on). Therefore, a minimum voltage of 4.0V is required to operate continuously within specification. A minimum of 3.9V (typical) is also required for startup."

e) In the Electrical Characteristics Tables, it says that the UVLO (under voltage lockout) threshold, for a rising input, can be as high as 4.2V (3.9V typical). It also says that as input falls, the threshold has a max of 3.2V (2.9V typical).

All this should make it exceedingly clear to you that *any statement* outside of the section *explicitly* titled "Electrical Characteristics," in fact even *one* line above it or below it, is probably just a "typical" (*not guaranteed*) value. And most engineers know that "typical" doesn't really mean a thing in an actual design (except maybe to set the voltage divider ratio). For example, the electrical tables *guarantee* that you will start up at 4.2V (not less). So if your circuit was intended to start up at 4V (based on the cover page), you may even get a whole bunch of boards working on your bench. But when you proceed towards mass production, the story will sour quickly. And if you go back to the company that sold you the part, you can bet their legal department will show you all the fine print you missed.

On page 9 of the datasheet, you also realize that in "sleep mode," the converter is basically operating as a hysteretic switcher—exhibiting bursts of pulses and then a long waiting time before the next burst. And the I_Q of 40μA refers only to the current drawn *between* the bursts of pulsing, that is, *during the periods of complete inactivity*. However relevant or meaningful that may be to the user.

I have come to believe recently, that the arena where the battle of specmanship is being currently waged big time *is the "I_Q spec."* Everybody seems to be doing their best to somehow state a lower number. They are betting that most engineers won't look too closely at what *their specific definition* of I_Q really is. Look closely.

Question 14

March 2006   Is it possible to connect the 2679's current adjust pin to GND without damage to the device? In my design the 2679 must deliver 5A at 27V with a Vin of 36V (duty cycle 75%). During testing, I saw that above 3.2A, the 2679 limits the duty cycle and the output voltage drops to 16V. The only way to maintain the correct output voltage is reduce the current limit resistor to a lower value (<1 kOhm). Though the recommended value in the datasheet for that is >5.6 kOhm.

Blog Entry 1   It's probably not a good idea to set the current limit to infinity (shorting Pin 5 to GND), because the 2679 will not be protected from short-circuit condition. Hence it will pose a safety hazard to both the equipment and personnel. In your application, there is a high probability that the 2679 may be going into thermal shutdown because it is being operated at high power dissipation. Based on our virtual bench recommendations, you should use the 2679T-ADJ, mounted on a heatsink rated at 4.8°C/W, e.g., Aavid P/N 532702B2500. This solution is only good for an ambient temperature of 30°C or lower.

Blog Entry 2 What inductor are you using? You could in fact be in current limit by using too low an inductor value, or saturating the inductor. I would measure your inductor current and make sure peaks are <5.75A. Maybe check the design on virtual bench too. After all you're trying to deliver >125W.

End of Thread.

Author's Comments: Judging by previous datasheet revision dates and various other blogs, the company actually knew about the D > 0.5 problem of their 267x family since early 2003. But three years later, their officials on the forum seem to imply they still don't know what this customer may be talking about here. The customer has said the duty cycle of the application is greater than 75% and also that "the 2679 limits the duty cycle," and that "the output voltage drops to 16V." Apparently that level of detail is just not enough for the company official!

One correct piece of advice though is *not* to try and increase the current limit of the switcher (that's an easy "fix" to the D > 0.5 problem—remove the offending current limit circuit altogether). The switch is clearly not rated for such high currents, nor has the company obviously evaluated the reliability under such conditions. So that warning is certainly fair and due. What's wrong is all the rest:

a) How can it be a thermal problem when the device is not even able to deliver the required output? The customer has also not implied that the observed problem only occurs after some time (after heating up). And surely, if overtemperature protection engages, the switch will be turned OFF completely for some time—so why would there be a 16V output still remaining at the output? Is the company official also saying that this particular heatsink alone (with its impressive part number) is the only solution (perhaps at subzero temperatures)? Then he or she also mentions "virtual bench." But in fact their online tool still doesn't warn you that you are in a D > 0.5 situation where you may need to be aware of certain issues.

b) The company official warns that the inductor may be saturating. Well, all switchers *can* momentarily saturate their inductors at power-up. That has never posed a problem in low-voltage applications. For example, if you are using a 5A Buck switcher for a 2A application, even the IC vendor usually suggests in the accompanying application guidelines that you choose an inductor with a saturation current rating of about 2.5A. In that case, if you then look at the inductor current on a scope at power-up, you will almost certainly see it hitting around 5.5 to 7A—because that's where the current limit of a 5A switcher would be (i.e., anywhere within its "ICLIM" MIN to MAX range). Clearly, a 2.5A inductor will saturate at 5.5A. But nothing really happens, because the switch can always protect itself (usually).

c) And then he says "After all you're *trying* to deliver > 125W." Not exactly. The customer just said that at 3.2A the output folds back to 16V. That's only 50W. I bet at that moment the customer is simply trying to get up to 60W. And to put it in perspective, any 12V/5A switcher regularly delivers 60W without even whimpering. For example, the LTC3780 is a *4-switch* non-inverting Buck-Boost IC that delivers 12V/5A from a PCB about 1.5 × 1.5in (2.5 in²) in size. Whereas the 2679 eval board seems to be about 5 × 3in (15 in²), judging by its pictures on the web.

Question 15

February 2006   Don't miss page 12 of the DS of 2678 for example. Check if it applies to you. Tucked away is "public" information that you were likely to miss (expectedly!!!). In fact here is a whole series of parts that can't function properly if the duty cycle is greater than 50%. See their datasheet, but a little more closely from now on!! On page 12 of the 2678 datasheet:

ADDITIONAL APPLICATION INFORMATION: When the output voltage is greater than approximately 6V, and the duty cycle at minimum input voltage is greater than approximately 50%, the designer should exercise caution in selection of the output filter components. When an application designed to these specific operating conditions is subjected to a current limit fault condition, it may be possible to observe a large hysteresis in the current limit. This can affect the output voltage of the device until the load current is reduced sufficiently to allow the current limit protection circuit to reset itself. Under current limiting conditions, the 267x is designed to respond in the following manner: At the moment when the inductor current reaches the current limit threshold, the ON-pulse is immediately terminated. This happens for any application condition. However, the current limit block is also designed to momentarily reduce the duty cycle to below 50% to avoid subharmonic oscillations, which could cause the inductor to saturate. Thereafter, once the inductor current falls below the current limit threshold, there is a small relaxation time during which the duty cycle progressively rises back above 50% to the value required to achieve regulation. If the output capacitance is sufficiently large, it may be possible that as the output tries to recover, the output capacitor charging current is large enough to repeatedly retrigger the current limit circuit before the output has fully settled. This condition is exacerbated with higher output voltage settings because the energy requirement of the output capacitor varies as the square of the output voltage (1/2CV2), thus requiring an increased charging current. A simple test to determine if this condition might exist for a suspect application is to apply a short circuit across the output of the converter, and then remove the shorted output condition. In an application with properly selected external components, the output will recover smoothly. Practical values of external components that have been experimentally found to work well under these specific operating conditions are COUT = 47uF, L = 22uH. It should be noted that even with these components, for a device's current limit of ICLIM, the maximum load current under which the possibility of the large current limit hysteresis can be minimized is ICLIM/2. For example, if the input is 24V and the set output voltage is 18V, then for a desired maximum current of 1.5A, the current limit of the chosen switcher must be confirmed to be at least 3A.

Blog Entry 1   The condition described is the result of what amounts to a foldback current limit design that's intended to prevent damage to either the regulator or the load under unusual fault conditions. Anyone familiar with foldback current limit will realize that there are always conditions that can be realized that force the foldback to get "stuck" in a stable, low output voltage operating mode. The solution, in general, is to reduce the load until the output is allowed to recover. The datasheet clearly advises the user what to look for and how to deal with any potential problems that may arise from this. Any implication that the information is intentionally obscured is clearly misleading on the part of the author.

End of Thread.

> *Author's Comments:* I have talked about this in Chapter 8. Basically what the very long explanation (over 400 words) says is that this *5A switcher can work only up to 2.5A* if your application requires an output greater than half the input. And that too, only with the specific C_{OUT} and L listed. My first question is, why would anyone *deliberately* design a switcher IC in that way (as the responder suggests)? My second question is, if the responder's response was truly honest, why wasn't the innocent explanation put forward just one month later to the customer in Question 14? Or less than a month earlier—to the customer in Question 16 below? Oh, the responder seems to be their new Apps manager. First time on this forum in years.

Question 16

February 2006   I am using the 2678 as an adjustable Buck regulator and am having difficulties with the voltage output. It "sets" to the proper value using calculated values for resistors, but upon applying a small (<1A) load, the voltage will dip up to 1.5V lower. Which components might we focus our debugging efforts on? The Buck regulator is a standalone PCB with no other circuitry operated concurrently. Any suggestions are welcome.

The following is a parts list and a schematic.

> IC: 2678S-ADJ-ND regulator 5A/30V
>
> EEU-FM1E221 Cout—220uF, electrolytic
>
> EEU-FM1H101 Cin—100uF, electrolytic
>
> C0603C103J5RACTU Cboost—0.01uF, ceramic
>
> CDRH127-330MC L 22uH
>
> 90SQ040 D1,D2 35V/9A (Schottky)
>
> P40.2KAACT-ND R2—40.2k 1%
>
> P7.87KAACT-ND R1—7.87k 1%

Blog Entry 1   Please tell us the rest of the "story." What is the input voltage range? What load current is used in the circuit design? I need this data to verify the parts you listed. I noted that the part number of the inductor indicates that it is a 33uH part, not 22uH as you indicated. Another part you should try to check is the output capacitor. Make sure that its current rating is equal to or greater than the peak-to-peak inductor ripple current or roughly about 30% of the Iout(max). The output cap's role is not to be taken for granted, because it discharges stored energy into the load (along with the inductor) during the OFF time of the switch.

Author's Comments: So using a 33μH instead of 22μH will create all sorts of problems, huh? Well maybe in this particular case that is true, because from Question 15 it seems the (only) recommended value for this switcher is 22μH—for any load current, if the duty cycle exceeds 50%.

I am also puzzled why the RMS current rating of the output capacitor becomes an issue in even getting the output to come up the *first time*. Sure, the RMS is important, but only for meeting long-term reliability and life objectives. But I have never seen a case where the capacitor loudly complains of abuse and quits working altogether. But if it is being implied that the energy in the capacitor is *not enough*, you have to understand that the customer is using a 220μF capacitor, whereas they have recommended a max of 47μF in the datasheet (see Question 15).

I decided to also check the statement that the current rating of the capacitor needs to be 30% of I_O. In Figure 12-14 I have carried out a hand derivation of the general equation for output RMS. We thus get

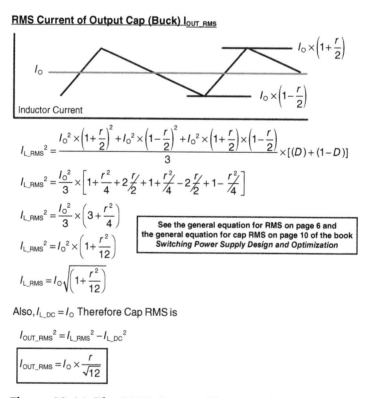

RMS Current of Output Cap (Buck) I_{OUT_RMS}

$I_O \times \left(1 + \dfrac{r}{2}\right)$

I_O

$I_O \times \left(1 - \dfrac{r}{2}\right)$

Inductor Current

$$I_{L\_RMS}^2 = \frac{I_O^2 \times \left(1+\frac{r}{2}\right)^2 + I_O^2 \times \left(1-\frac{r}{2}\right)^2 + I_O^2 \times \left(1+\frac{r}{2}\right) \times \left(1-\frac{r}{2}\right)}{3} \times [(D) + (1-D)]$$

$$I_{L\_RMS}^2 = \frac{I_O^2}{3} \times \left[1 + \frac{r^2}{4} + 2\frac{r}{2} + 1 + \frac{r^2}{4} - 2\frac{r}{2} + 1 - \frac{r^2}{4}\right]$$

$$I_{L\_RMS}^2 = \frac{I_O^2}{3} \times \left(3 + \frac{r^2}{4}\right)$$

$$I_{L\_RMS}^2 = I_O^2 \times \left(1 + \frac{r^2}{12}\right)$$

> See the general equation for RMS on page 6 and the general equation for cap RMS on page 10 of the book *Switching Power Supply Design and Optimization*

$$I_{L\_RMS} = I_O \sqrt{\left(1 + \frac{r^2}{12}\right)}$$

Also, $I_{L\_DC} = I_O$ Therefore Cap RMS is

$$I_{OUT\_RMS}^2 = I_{L\_RMS}^2 - I_{L\_DC}^2$$

$$\boxed{I_{OUT\_RMS} = I_O \times \frac{r}{\sqrt{12}}}$$

Figure 12-14 The RMS Current Through the Output Capacitor of a Buck

$$C_{OUT} \times I_{RMS} \times I_{OUT\_RMS} = I_O \times \frac{r}{\sqrt{12}} \approx 30\% \times I_O \times r$$

So for a typical r of 0.4, we get about 12%, not 30%, of I_O.

Blog Entry 2   I apologize for the lack of clarity. The inductor is a 33uH. When I designed the circuit, it was with a max current load of 5A in mind. The thought was that if it was spec'd to this value, it would work fine for the current application, which draws about 1A on average. The input voltage was designed with the range of 12V–36V. Currently, we are supplying it with a 15V battery. The output voltage is desired to be 7.4V. I will check the current rating on the capacitor and inductor. Thanks for the help.

> *Author's Comments:* The customer may have thought or assumed that "if it was spec'd to this value, it would work fine for the current application," but he or she actually needed to confirm that from the vendor. We should always be wary of implied expectations, as pointed out previously on several occasions.
>
> The duty cycle is uncomfortably close to 50%. From the forum and datasheet, it seems that if this part hits its current limit even once, it folds back to about 40 to 45% duty cycle for a rather long time. And it may never rise after that ("motorboating"). So a 15V input *could* bring the output down to about 6V output momentarily, close to what the customer is complaining about. It may not be very likely, but it is a plausible explanation at this stage, and should be discussed with the customer.
>
> Having got the customer to apologize for a minor typo (instead of the company apologizing for what is more likely a major screw-up in design), let us see if they at least guide him correctly from this point on.

Blog Entry 3   The capacitor's ripple current is rated at 950mA at 100kHz, which falls within our application spec, I believe. Its impedance at 100kHz is given as 0.056Ohms.

Blog Entry 4   The maximum current drawn from the supply with a 1.0A load current is about 1.2A. You mentioned that the input source is drawn from a 15V battery. Did you verify if the battery is adequately sized for this operational condition? Does the input voltage stay fixed at 15V when the 1.0A load is applied?

> *Author's Comments:* In going from 15V to 7.4V, the duty cycle is 50%. So in fact the input current with a 1A load is only 0.5A. This simple fact eludes many engineers, which is why I talked about it in so much detail in Chapter 2 (the "missing current" problem).

Blog Entry 5   I have just tested the battery and it maintains its voltage when the circuit has a 3A load. The battery itself is rated for many more amps than that.

Blog Entry 6   With the switching regulator putting out 3A using a resistive load, the voltage drops from 7.4V to around 4V (voltage is unsteady at this point). I can't seem to figure out what would be causing this. Thanks for your continued assistance.

Author's Comments: We got some new information from the customer. It could change our theory. We must learn never to *focus only on a single pet theory of ours. There may be other causes, or even multiple reasons for the same apparent symptom.* I now suspect there is apparently something else seriously wrong here. If it were only the $D > 0.5$ issue explained previously, the output would just fall a volt or so below the set 7.4V. So at this stage at least, I would start questioning the PCB layout, as well as the input decoupling.

Blog Entry 7 In reviewing your parts list, I noted that the inductor may be slightly underrated. With a 3A load, the current flowing through it will be approximately 3.45A, which is in excess of the inductor's 3A saturation current spec. So, the inductor is probably going into saturation (it turns into a resistor), thus causing Vout to drop. See if you can replace this with a DR127-220, which is rated 22.9uH, 4.0A DC max.

Author's Comments: That's obviously not true. The customer has had problems at 1A load too. So this explanation, even if true (and it is not) does not really fit the set of clues provided by the customer. It could be *one* reason, but is not *the* explanation. As indicated previously, the saturation rating of most inductors is not a cliff over which you suddenly fall off into a timeless abyss. "Saturation" usually just means the inductance has fallen by about 20 to 30% from its initial value. So in going further, say from 3A to 3.5A, you can expect the inductance to fall to about 50% of its initial value. But rarely any more! For sure, it doesn't suddenly morph into a "resistor."

Blog Entry 8 Thanks! I will order that part, give it a shot and post back with the results.

Blog Entry 9 The actual device we are powering draws a peak of 1.2A and averages about 0.25A. The voltage droops from a nominal 7.4V by about 0.5V during the peak current draw period. Adding an extra inductor in parallel (56uH/2.25A) with the original (33uH/3A) inductor seems to help by reducing the amount of voltage dropped to 0.2V. This does not make sense though, as the inductor already on the PCB is more than large enough for a peak draw of 1.2A. Is it possible that a component is broken? I have made many of these boards (and breadboards) and they all behave in the exact same way, so I suspect it is component selection causing the problem. Thanks for the help; I'm sure you are as tired of this problem as I am.

Author's Comments: That is again commensurate with the original $D > 0.5$ theory. By paralleling an extra inductor, the customer has reduced the effective inductance—closer to the 22µH suggested on page 12 of the datasheet. So why not suggest the customer simply try a single 22µH and be done with it? Why beat around the bush?

Blog Entry 10 I don't suspect that you have a failed component problem here. The most likely "culprit" in many PWM switching regulator problems is a bad layout design, which can cause a proliferation of stray capacitances and/or inductances in the assembly. I noted that you mentioned the word "breadboard," which is a "no-no" when you are trying to build a switching regulator. For good layout design, I'd like to refer you to the material I attached below. Also, you may try adding a small bypass capacitor, 0.1uF ceramic type, as close to the Vin pin as possible. You can also add another bypass capacitor in parallel with the feedback resistor (this is R2 in your schematic).

Blog Entry 11   I will try using the bypass capacitors. I think you are right—Layout is an art form I hear.

End of Thread.

> *Author's Comments:* That is finally the very first question that should have been asked of the customer—the PCB layout. But note that the symptoms also fit the $D > 0.5$ issue (and certainly will, when the customer goes to his proposed V_{INMIN} of 12V). So the customer should have been directed to page 12 right away. That never happened.
>
> Incidentally, don't blindly add a "bypass capacitor in parallel with the (upper) feedback resistor," as suggested. That feedforward capacitor introduces another zero in the loop and can cause the system to go unstable. You should realize that this family of devices has a full-blown internal Type 3 compensation, so it even has an internal zero to emulate an external "ESR zero." That is why this family is supposed to be able to handle ceramic capacitors at the output. If you introduce yet another zero (via the feedforward capacitor as suggested), you could have one too many zeros. And ultimately, your design could be one, too (a zero).
>
> Note how some customers are just too embarrassed to return and publicly declare "it doesn't work." They silently blame their own lack of skills or their "bad" PCB layout and move on. They assume this huge and famous company couldn't be all that wrong! But how very convenient that assumption is for some, they really have no idea. It turns out that "thinking engineers" was actually the last thing they ever had in mind!

Appendix

	Buck	Boost	Buck-Boost
Duty Cycle	$\dfrac{V_O + V_D}{V_{IN} - V_{SW} + V_D}$	$\dfrac{V_O - V_{IN} + V_D}{V_O - V_{SW} + V_D}$	$\dfrac{V_O + V_D}{V_{IN} + V_O - V_{SW} + V_D}$
V_{IN_50} (V)	$(2 \cdot V_O) + V_{SW} + V_D$ $\approx 2 \cdot V_O$	$\dfrac{1}{2} \cdot [V_O + V_{SW} + V_D] \approx \dfrac{V_O}{2}$	$V_O + V_{SW} + V_D \approx V_O$
Output Voltage, V_O (V)	$V_{IN} \cdot D - V_{SW} \cdot$ $D - V_D \cdot (1 - D)$	$\dfrac{V_{IN} - V_{SW} \cdot D - V_D \cdot (1 - D)}{1 - D}$	$\dfrac{V_{IN} \cdot D - V_{SW} \cdot D - V_D \cdot (1 - D)}{1 - D}$
Voltµseconds (Vµs)	$\dfrac{V_O + V_D}{f} \cdot (1 - D) \cdot 10^6$	$\dfrac{V_O - V_{SW} + V_D}{f} \cdot D \cdot (1 - D) \cdot 10^6$	$\dfrac{V_O + V_D}{f} \cdot (1 - D) \cdot 10^6$
L (µH)	$\dfrac{V_O + V_D}{I_O \cdot r \cdot f} \cdot (1 - D) \cdot 10^6$	$\dfrac{V_O - V_{SW} + V_D}{I_O \cdot r \cdot f} \cdot D \cdot (1 - D)^2 \cdot 10^6$	$\dfrac{V_O + V_D}{I_O \cdot r \cdot f} \cdot (1 - D)^2 \cdot 10^6$
Inductor Current Ripple Ratio r	$\dfrac{V_O + V_D}{I_O \cdot L \cdot f} \cdot (1 - D) \cdot 10^6$	$\dfrac{V_O - V_{SW} + V_D}{I_O \cdot L \cdot f} \cdot D \cdot (1 - D)^2 \cdot 10^6$	$\dfrac{V_O + V_D}{I_O \cdot L \cdot f} \cdot (1 - D)^2 \cdot 10^6$
ΔI (A)	$\dfrac{V_O + V_D}{L \cdot f} \cdot (1 - D) \cdot 10^6$	$\dfrac{V_O - V_{SW} + V_D}{L \cdot f} \cdot D \cdot (1 - D) \cdot 10^6$	$\dfrac{V_O + V_D}{L \cdot f} \cdot (1 - D) \cdot 10^6$
RMS Current in Input Capacitor (A)	$I_O \cdot \sqrt{D \cdot \left[1 - D + \dfrac{r^2}{12}\right]}$	$\dfrac{I_O}{1 - D} \cdot \dfrac{r}{\sqrt{12}}$	$\dfrac{I_O}{1 - D} \cdot \sqrt{D \cdot \left[1 - D + \dfrac{r^2}{12}\right]}$
I_{PP} in Input Capacitor (A)	$I_O \cdot \left[1 + \dfrac{r}{2}\right]$	$\dfrac{I_O \cdot r}{1 - D}$	$\dfrac{I_O}{1 - D} \cdot \left[1 + \dfrac{r}{2}\right]$
RMS Current in Output Capacitor (A)	$I_O \cdot \dfrac{r}{\sqrt{12}}$	$I_O \cdot \sqrt{\dfrac{D + \dfrac{r^2}{12}}{1 - D}}$	$I_O \cdot \sqrt{\dfrac{D + \dfrac{r^2}{12}}{1 - D}}$
I_{PP} in Output Capacitor (A)	$I_O \cdot r$	$\dfrac{I_O}{1 - D} \cdot \left[1 + \dfrac{r}{2}\right]$	$\dfrac{I_O}{1 - D} \cdot \left[1 + \dfrac{r}{2}\right]$

	Buck	Boost	Buck-Boost
Energy Handling Capability (μJoules)	$\dfrac{I_O \cdot V\mu s}{8} \cdot \left[r \cdot \left(\dfrac{2}{r} + 1 \right)^2 \right]$	$\dfrac{I_O \cdot V\mu s}{8 \cdot (1-D)} \cdot \left[r \cdot \left(\dfrac{2}{r} + 1 \right)^2 \right]$	$\dfrac{I_O \cdot V\mu s}{8 \cdot (1-D)} \cdot \left[r \cdot \left(\dfrac{2}{r} + 1 \right)^2 \right]$
RMS Current in Inductor (A)	$I_O \cdot \sqrt{1 + \dfrac{r^2}{12}}$	$\dfrac{I_O}{1-D} \cdot \sqrt{1 + \dfrac{r^2}{12}}$	$\dfrac{I_O}{1-D} \cdot \sqrt{1 + \dfrac{r^2}{12}}$
Average Current in Inductor (A)	I_O	$\dfrac{I_O}{1-D}$	$\dfrac{I_O}{1-D}$
RMS Current in Switch (A)	$I_O \cdot \sqrt{D \cdot \left[1 + \dfrac{r^2}{12} \right]}$	$\dfrac{I_O}{1-D} \cdot \sqrt{D \cdot \left[1 + \dfrac{r^2}{12} \right]}$	$\dfrac{I_O}{1-D} \cdot \sqrt{D \cdot \left[1 + \dfrac{r^2}{12} \right]}$
Peak Current Switch/Diode/ Inductor (A)	$I_O \cdot \left[1 + \dfrac{r}{2} \right]$	$\dfrac{I_O}{1-D} \cdot \left[1 + \dfrac{r}{2} \right]$	$\dfrac{I_O}{1-D} \cdot \left[1 + \dfrac{r}{2} \right]$
Average Current in Switch (A)	$I_O \cdot D$	$I_O \cdot \dfrac{D}{1-D}$	$I_O \cdot \dfrac{D}{1-D}$
Average Current in Diode (A)	$I_O \cdot (1-D)$	I_O	I_O
Average input Current (A)	$I_O \cdot D$	$\dfrac{I_O}{1-D}$	$I_O \cdot \dfrac{D}{1-D}$
Output Voltage Ripple ($\pm$mV)*	$\dfrac{1}{2} \cdot I_O \cdot r \cdot ESR(m\Omega)$	$\dfrac{1}{2} \cdot \dfrac{I_O}{1-D} \cdot \left[1 + \dfrac{r}{2} \right] \cdot ESR(m\Omega)$	$\dfrac{1}{2} \cdot \dfrac{I_O}{1-D} \cdot \left[1 + \dfrac{r}{2} \right] \cdot ESR(m\Omega)$

$r = \Delta I / I_{DC}$, L in μH, f in Hz. All voltages and currents are magnitudes. * ESL ignored.

Index

Printed and bound by CPI Group (UK) Ltd, Croydon, CR0 4YY

03/10/2024

01040335-0003